Optical Near Fields

Advanced Texts in Physics

This program of advanced texts covers a broad spectrum of topics which are of current and emerging interest in physics. Each book provides a comprehensive and yet accessible introduction to a field at the forefront of modern research. As such, these texts are intended for senior undergraduate and graduate students at the MS and PhD level; however, research scientists seeking an introduction to particular areas of physics will also benefit from the titles in this collection.

Springer
Berlin
Heidelberg
New York
Hong Kong
London
Milan
Paris
Tokyo

M. Ohtsu K. Kobayashi

Optical Near Fields

Introduction
to Classical and Quantum Theories
of Electromagnetic Phenomena
at the Nanoscale

With 108 Figures, 20 Problems and Solutions

Springer

Professor Motoichi Ohtsu
Tokyo Institute of Technology
4259 Nagatsuta-cho, Midori-ku
Yokohama 226-8502, Japan
E-mail: ohtsu@ae.titech.ac.jp

Dr. Kiyoshi Kobayashi
ERATO Localized Photon Project
JST, 4th Floor, Tenko Building No. 17
687-1 Tsuruma, Machida
Tokyo 194-0004, Japan
E-mail: kkoba@ohtsu.jst.go.jp

Library of Congress Cataloging-in-Publication Data: Ohtsu, Motoichi. Optical near fields: electromagnetic phenomena at the nanoscale/ M. Ohtsu, K. Kobayashi. p. cm. – (Advanced texts in physics, ISSN 1439-2674) Includes bibliographical references and index. (alk. paper) 1. Photonics. 2. Nanotechnology. 3. Quantum optics. 4. Near-field microscopy. I. Kobayashi, K. (Kiyoshi), 1953– II. Title. III. Series. TA1520.O38 2003 621.36–dc21 2003054286

ISSN 1439-2674

ISBN 978-3-642-07343-4

Springer-Verlag is a part of Springer Science+Business Media

springeronline.com

Printed in Germany

Final layout: Frank Herweg, Leutershausen
Cover design: *design & production* GmbH, Heidelberg

Printed on acid-free paper

Preface

This book outlines physically intuitive concepts and theories for students, engineers, and scientists who will be engaged in research in nanophotonics and atom photonics. The main topic is the optical near field, i.e., the thin film of light that is localized on the surface of a nanometric material. In the early 1980s, one of the authors (M. Ohtsu) started his pioneering research on optical near fields because he judged that nanometer-sized light would be required to shift the paradigm of optical science and technology. This field of research did not exist previously, and was not compatible with trends in optical science and technology. However, he was encouraged by the knowledge that scientists in other countries started similar research in the mid 1980s.

In the 1990s, optical technology progressed very rapidly and the photonics industry developed, but further progress became difficult due to the fundamental limit of light known as the diffraction limit. However, there was a growing awareness among scientists and engineers that this limit can be overcome using optical near fields. Since a drastic paradigm shift in the concepts of optics is required to understand the intrinsic nature of optical near fields, the demand for a textbook on this subject has increased. The present book aims to meet this demand.

Most scientists and engineers believe that optical near fields can only be applied to microscopy. Therefore, this field of applications has been called near-field optics. However, applications to microscopy never exploit the essential nature of optical near fields. Although this book discusses the application to microscopy as a simple topic in order to guide beginners in the study of optical near fields, the main purpose here is to justify a much wider claim, i.e., that the essence of utilizing optical near fields is to realize novel nanometric processing, function, and manipulation by controlling an intrinsic interaction between nanometer-sized optical near fields and material systems or atoms. This has not been realized by conventional optical science and technology. M.O. refers to the novel optical science and technology in nanometric and atomic regions as nanophotonics and atom photonics, respectively. As long as this more fundamental aspect exists, realizing nanometer-sized optical science technology beyond the diffraction limit is no more than a by-product.

The book neither reviews formulae for numerical calculations nor introduces experimental results on microscopy. It describes physically intuitive theories in three separate parts. Appendices A–D provide supplementary explanations.

Chapters 1–3 review the background, history, and present status of research and development in optical near fields. Chapter 1 reviews the history of optics and early progress in photonics, i.e., applications of lasers to optical disk memory, optical fiber communication, and optical microfabrication. It shows how progress in photonics reached a deadlock due to the diffraction limit. Chapter 2 describes how to end this deadlock and realize nanophotonics for the requirements of the 21st century. This is possible by generating and utilizing optical near fields and the principles for doing so are described in this chapter. These reviews introduce the theoretical discussions later in the book. Chapter 3 discusses the practical aspects of the novel science and technology related to optical near fields. It also describes the structure and performance of probes, which are key devices for generating or detecting optical near fields. The present status of research and development is reviewed, including applications to microscopy, spectroscopy, fabrication, optical disk memory, and atom manipulation. Finally, this chapter overviews possible trends in the novel fields of optical science that will be founded by exploiting optical near fields.

Chapters 4–8 provide the theoretical basis for optical near fields. They describe a theoretical model for the electromagnetic interaction between two or more nanometric material systems located in proximity. Although one can derive some information on optical near fields by simultaneously solving the approximated Maxwell and Schrödinger equations, this derivation requires numerical calculations with a very long computation time. Although numerical results are possible, it is very difficult to obtain an intuitive physical picture of the physics of optical near fields. In order to overcome this difficulty, these four chapters are devoted to reviewing theoretical models that offer intuitive concepts for analyzing the physical meaning of optical near fields and relevant experimental results.

Chapter 4 presents the simplest theoretical model to describe the phenomena presented in Chap. 2. It imposes a condition on the size of the material systems in which optical near fields are investigated. Under this condition, the basic role of a probe is described from the viewpoint of a dipole–dipole interaction. The characteristics of fiber probes, which depend on their shape and composition, are also discussed.

Chapter 5 deals with a single atom or molecule as a nanometric material system to investigate its basic spectral properties. If a conducting or dielectric probe approaches the nanomaterial, the emission properties of the atom or molecule are substantially modified. This phenomenon is discussed by dealing with the atom or molecule as an oscillating electric dipole moment. After presenting the basic concepts, an analytical method is given in which the probe tip is approximated as a planar mirror. The results of a quantum mechanical approach are also described.

Chapter 6 discusses the effect of multiple scattering for the more precise investigation of optical near fields. A propagator, i.e., the transfer function, is derived, in order to evaluate the electric field at an arbitrary position generated by a light source at another position. The result of this derivation is applied to near-field optical microscopy.

In Chap. 7, by introducing a dual vector potential and a scalar potential, the basic formulae of electromagnetics are transformed and a novel theoretical model is presented. This model is put to use for a systematic analysis of several cases. In the first case, the near-field condition is met, i.e., the sizes of the material systems under study and their separation are sufficiently smaller than the wavelength of the incident light. The second case is the quasi-near-field condition, i.e., the near-field condition is not met with sufficiently high accuracy.

Chapter 8 presents a quantum mechanical model and a new approach based on a projection operator method to describe the interaction between nanometric material systems via optical near fields. This model can also be used to describe the interaction between an atom and a probe, and its application to atom photonics is discussed in the last two chapters of the book. An outstanding advantage of this model is its ability to describe systematically the light–matter interactions in nanometric material and atomic systems. This is because the model is based on concepts developed in the fields of elementary particle physics, statistical mechanics, quantum chemistry, and quantum optics. Furthermore, the model provides an intuitive physical picture in which the localized optical near fields can be described as an electron cloud localized around an atomic nucleus.

Utilizing the theoretical basis presented in Chaps. 4–8, Chap. 9 discusses the possibility of creating new fields in nanophotonics and atom photonics, to shift the paradigm of optical science and technology.

Chapters 1–7, and 9 were written by M. Ohtsu, whilst Chaps. 8 and 9 were by K. Kobayashi. Both authors checked the whole manuscript. The authors thank Drs. H. Hori, I. Banno (Yamanashi University), Drs. T. Saiki, S. Mononobe, R. Uma Maheswari, K. Kurihara, M. Ashino, M. Naya, J.D. White, K. Matsuda, N. Hosaka (Kanagawa Academy of Science and Technology), Drs. G.H. Lee, V. Polonski, T. Yatsui, T. Kawazoe, T.W. Kim, H. Aiyer, S. Sangu, K. Totsuka, S.M. Iftiquar, A. Shojiguchi (Japan Science and Technology, Corp.), Drs. H. Ito, M. Kourogi, A. Zvyagin, H. Fukuda, S.J. Lee, Y. Yamamoto, and H. Takamizawa (Tokyo Institute of Technology) for their collaboration in conducting the research on nanophotonics and atom photonics, and preparing the manuscript for the book. They also extend special thanks to Drs. T.W. Kim, S. Sangu, T. Yatsui, S.J. Lee, and H. Aiyer for their critical readings and comments on the manuscript.

They gratefully acknowledge Dr. T. Asakura, editor of the Springer Series in Optical Sciences and Professor Emeritus at Hokkaido University, who recommended the publishing of this book. Finally, they wish to express their gratitude to Dr. C. Ascheron of Springer-Verlag, for his guidance and suggestions throughout the preparation of this book.

Yokohama, Kanagawa, *Motoichi Ohtsu*
Machida, Tokyo, *Kiyoshi Kobayashi*
September 2003

Contents

1 Deadlocks in Conventional Optical Science and Technology

The present chapter starts by reviewing the history of optics in Sect. 1.1. Then Sect. 1.2 describes early progress in photonics, i.e., applications of lasers to optical disk memory, optical fiber communication, and optical microfabrication. Finally, Sect. 1.3 shows how progress in photonics reached a deadlock due to the diffraction of light.

1.1 Progress in Optics

The history of optics is a long one. Even as early as the 17th century, Isaac Newton studied the nature of light and proposed the particle model of light [1.1]. This resulted in his long dispute with Robert Hooke and Christiaan Huygens who proposed the wave model of light. However, Newton also intensively studied the characteristics of an interference fringe called Newton's ring, showing that he also paid attention to the wave property of light. This implies that his main interest cannot have been to discuss whether the nature of light is a particle or a wave, but was rather to treat light as a particle and/or a wave depending on the property manifested. On the other hand, although Thomas Young supported the wave theory, he claimed that optical interference cannot prove the validity of the wave theory. These historical aspects suggest that both Newton and Young were aware of the fact that it was difficult to elucidate the particle model and the wave model separately.

Through these arguments over the nature of light described above, the idea arose that light might have both properties, those of a particle and those of a wave, i.e., the notion of the duality of light was born. Accordingly, light exhibits a particle-like nature through its energy, while it exhibits a wavelike nature through its phase. The problem is to discuss which property of light manifests itself, rather than to identify the nature of light. From these discussions, it was found that a quantum theory is required to describe the various aspects of the properties of light, including the problem of measurement.

Today, quantum theory claims that light has both the properties of a particle and the properties of a wave. Light is thus called a photon. When light behaves as a wave, it has a wavelength λ (period of spatial repetition) and a frequency ν (a number of oscillations per unit time). These are inversely proportional to each other, i.e., they are related by $\lambda = c/\nu$, where c is the speed of light. When light exhibits a particle-like nature, its energy is a multiple of $h\nu$, where h is Planck's constant with the value of 6.626176×10^{-34} J s.[1] Therefore, from the energy viewpoint, being a fundamental physical quantity, only the frequency ν is a significant quantity, even though λ is inversely proportional to ν.

Concerning the progress of quantum theory and its applications, a novel light source known as the laser was invented in 1960 [1.2]. The use of lasers has led to a dramatic change in optical science and technology, and its invention has been evaluated as one of the biggest scientific events in the twentieth century, on a par with the invention of the transistor. It can be claimed that the laser is an artificial light source invented by utilizing the accumulated scientific and technological resources of modern science and technology. The laser can be regarded as an ideal light oscillator with unique properties, completely different from those of a white light source or thermal radiation source such as the sun, a flash lamp, or a light bulb. It is the source of a sinusoidally oscillating electromagnetic wave, similar to conventional lower-frequency electronic oscillators. Laser light therefore exhibits far more remarkable properties than light emitted from conventional sources [1.3], i.e., high directivity, high spatial and temporal coherence, high power density, and high brightness. Further, its amplitude, phase, frequency, and polarization can be controlled and modulated very precisely. These considerable control and modulation capabilities are inherent advantages of the laser which have not been achieved by conventional light sources. Therefore, lasers have found a variety of applications.

1.2 Major Photonics Technologies and Their Limits

Industrial applications of lasers have been called photonics or opto-electronics. Optical disk memory and optical fiber communication systems are reviewed here as successful examples in photonics. Photolithography is also discussed, used to fabricate the photonic devices that make up these systems.

1.2.1 Optical Disk Memory System

A compact disk (CD), popularly used as a read-only memory (ROM), is an example of an optical disk memory which has a number of small pits on its surface to store digital signals, i.e., one pit corresponds to one bit. In order to read out these signals, the disk surface is illuminated by a laser beam which is focused by a convex lens, as shown in Fig. 1.1. Detection of laser light power reflected from the disk surface corresponds to the readout operation. The advantages of this disk system are long lifetime due to non-contact readout,

[1] The quantity $h/2\pi$ which is used in Chaps. 8 and 9, where it is represented by $\hbar$ and called Dirac's constant.

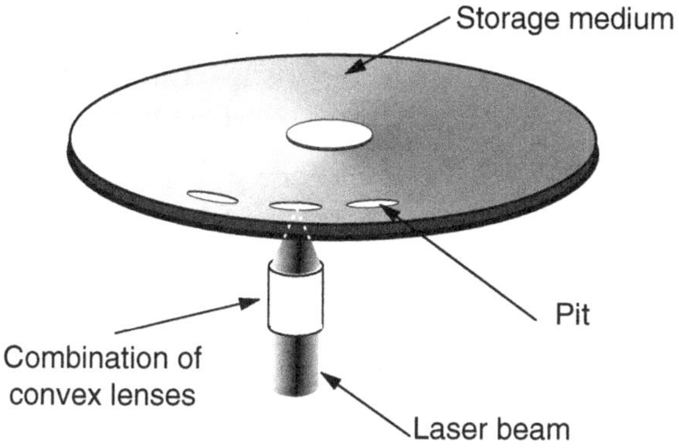

Fig. 1.1. Schematic diagram of an optical disk memory system

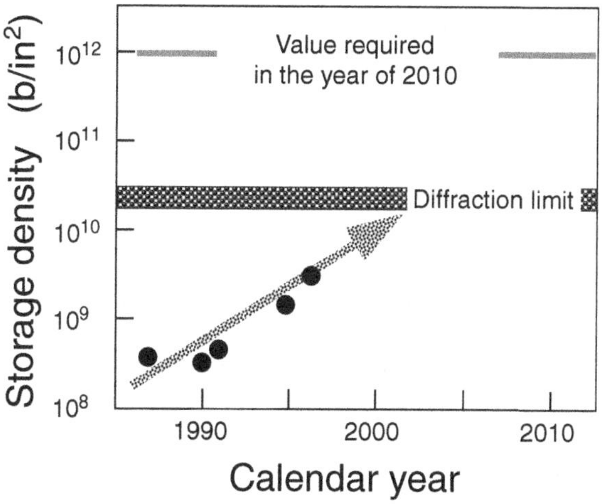

Fig. 1.2. Technical road map showing the increase in storage density of optical disk memory

high density due to the small spot size of the focused laser beam, and low noise due to digital signal processing. Random access memories (RAM), e.g., digital versatile disks (DVD), have also been developed, where a focused laser beam is also used to store and rewrite by heating the disk surface locally. Figure 1.2 shows progress in increasing the storage density of optical disk memory systems, at the annual increase rate of 30%.

A report on future trends in photonics technology has recently estimated the storage density and readout speed required of optical disk memory in the year 2010 [1.4], i.e., 1 Tb/in^2 and 100 Mb/s, respectively. However, the diameter of the circular pit corresponding to the 1 Tb/in^2 density is as small as 25 nm, which cannot be fabricated for the reasons to be explained in Sect. 1.3.

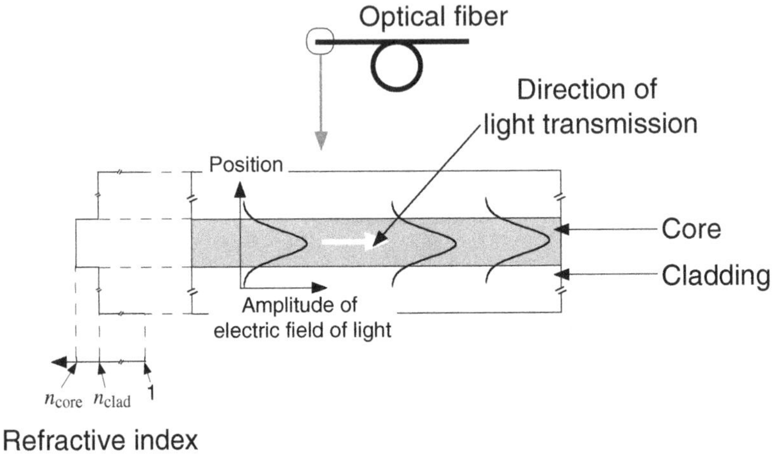

Fig. 1.3. Schematic diagram of an optical fiber. n_{core} and n_{clad} represent the refractive indices of the core and cladding, respectively

1.2.2 Optical Fiber Communication

An optical fiber is made from a coaxial silica glass with typical diameter 125 μm. Its inner and outer parts are called the core (about 3 μm in diameter) and the cladding, respectively, as shown in Fig. 1.3. The refractive index of the core is higher than that of the cladding and this means that they form an optical waveguide for light focusing and low-loss transmission. By dehydrating the silica glass, the transmission loss has been reduced to as low as 0.2 dB/km in the 1.5 μm wavelength region, which means that the transmitted light power decreases only 3.6% even after transmission as far as 1 km. As a result, long-distance optical transmission has become possible for the submarine optical communication cable system in the Pacific and Atlantic oceans.

The light sources used for optical fiber communication systems are semiconductor lasers with the structure shown in Fig. 1.4. Light is emitted from an active layer with typical thickness, width, and length 0.1 μm, 2 μm, and 300 μm, respectively. That is, by injecting current into this layer, light is emitted by the electronic interband transition from conduction to valence bands. The refractive index of the active layer is made higher than that of the adjacent cladding layers in order to form an optical waveguide for effective propagation of the emitted light. The light propagating through the active layer is reflected at the end facets for a round trip, i.e., the two end facets work as a cavity resonator. As a result of this round trip, the light is confined in the cavity for lasing [1.3]. A part of the lasing light leaks from the cavity and is used as the output laser light. The injection current is modulated in order to modulate the light power when carrying digital signals. To distribute the signals transmitted through the optical fiber to each

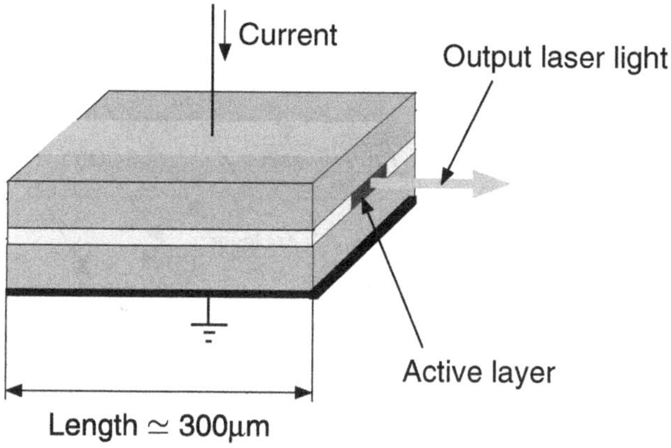

Fig. 1.4. Schematic diagram of a semiconductor laser

receiver, optical switching devices have been used. These are composed of optical waveguides, semiconductor lasers, and so on. Photonic integrated circuits have been developed to integrate these photonic devices on a common substrate.

A report on technological trends in optical fiber communication systems using photonics technology has recently estimated the dramatic increase in signal transmission rate that will be required in the year 2015 for domestic and local photonic networks [1.5]. For example, the numbers of input and output channels on an integrated optical switching array must be increased to as many as 3000 each in order to meet this requirement. This means that the size of each optical switching device must be reduced to the wavelength of light or even less in order to keep the integrated switching array as small as conventional arrays. However, this size reduction is not possible for the reason to be explained in Sect. 1.3.

1.2.3 Optical Microfabrication

Research and development of semiconductor microfabrication has been conducted with the support of national projects. It aims to fabricate ultralarge scale integrated circuits for semiconductor dynamic random access memories (DRAMs), photonic integrated circuits for optical fiber communication systems, and so on. Photolithography has been employed as a tool for microfabrication, using focused light to process the material surfaces. Fabricated sizes have been reduced rapidly using short wavelength light. For example, it has become possible to fabricate linear patterns of 100 nm width using an ArF eximar laser as light source (wavelength 196 nm). This will be used for mass production of 1 Gb or 4 Gb DRAMs in the near future.

It is estimated that 64–256 Gb DRAMs will be required in the early 21st century, as shown in Fig. 1.5. The linear pattern in these devices must be as

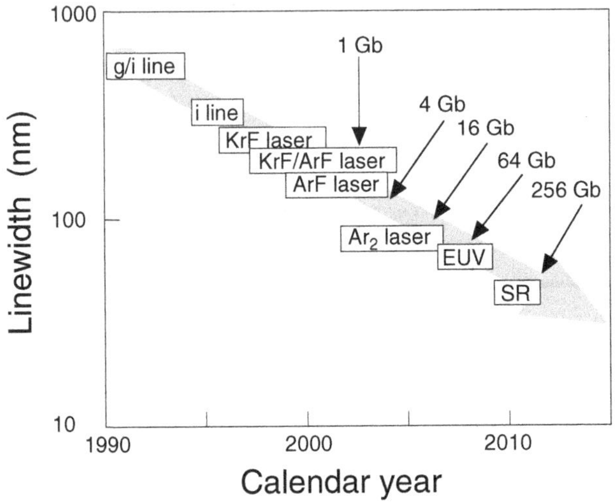

Fig. 1.5. Technical road map of the linewidth reduction of patterns fabricated by photolithography. Names of light sources and DRAM capacities are also shown. EUV and SR stand for extreme ultraviolet light and synchrotron radiation, respectively

narrow as 35–70 nm. However, such narrow patterns cannot be fabricated for the reason to be explained in Sect. 1.3. In order to make this possible, various light sources emitting extreme ultraviolet light, synchrotron radiation and X rays, as well as electron beams, are under development. However, they may not be feasible in mass production factories because of their large size and high cost. Semiconductor industries are expecting advances with novel, inexpensive, and practical fabrication tools.

1.3 Origin of Limits: Diffraction of Light

The three examples presented in Sect. 1.2 indicate that the society of the 21st century requires a novel optical technology in order to fulfill measurement, fabrication, control, and function requirements on a scale of several tens of nm. Such a technology can be called optical nanotechnology. However, conventional optical technology cannot meet this requirement. This is due to the diffraction limit of light waves.[2]

Figure 1.6 explains the phenomenon of diffraction schematically. A plane light wave propagates to a plate. After the light wave passes through a small aperture on the plate, it is converted to a diverging spherical wave. Such divergence is called diffraction [1.6]. It is an intrinsic characteristic of waves,

[2] For the formulation of diffraction, refer to Problems 1.1 and 1.2 given at the end of this chapter.

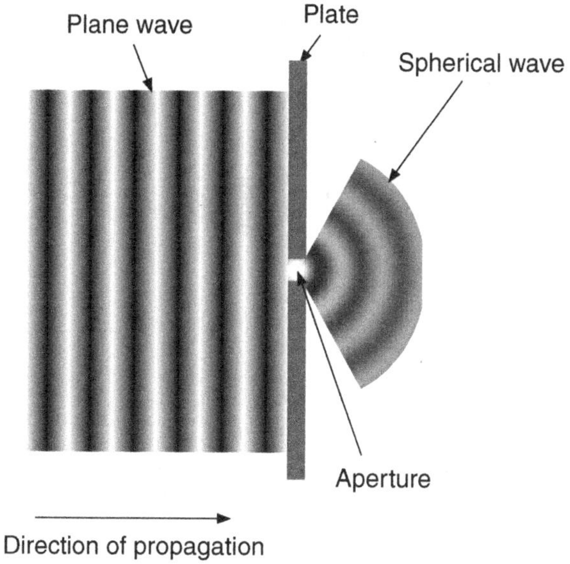

Fig. 1.6. Schematic representation of the diffraction phenomenon of light

exhibited by acoustic waves, ocean waves, and so on. In the case of a circular aperture, the divergence angle is about λ/a (radian), where λ and a are the wavelength of the incident light and the aperture radius, respectively.[3]

Due to this diffraction, the spot size of light cannot be zero even if it is focused by a convex lens. This is called defocusing. As shown in Fig. 1.7, the spot size on the focal plane is as large as λ/NA.[4] Here, the quantity NA is the numerical aperture given by $n \sin\theta$, where n is the refractive index of the medium between the lens and the screen, $\sin\theta = (a/2)/\sqrt{f^2 + (a/2)^2}$, a is the diameter of the circular convex lens, and f is the focal length. The value of NA is smaller than unity for conventional convex lenses.

Therefore, when two point sources of light are located closer together than λ/NA, their images formed by the convex lens cannot be resolved on the focal plane. This also holds true for imaging by an optical microscope. Thus, the smallest size resolvable by the optical microscope (i.e., the resolution) is λ/NA, which is called the diffraction limit. It is advantageous to use shorter wavelength light in order to improve the diffraction-limited resolution. Electron microscopes have realized very high diffraction-limited resolution because they utilize the short wavelength of the de Broglie wave of the electron. However, a disadvantage of the electron microscope is that the observable samples are limited to conductors installed in a vacuum, and this does not allow one to observe insulators and living biological samples.

[3] This expression is given by (Q1.9) in the solution to Problem 1.1.

[4] This expression is given by (Q1.10) in the solution to Problem 1.2.

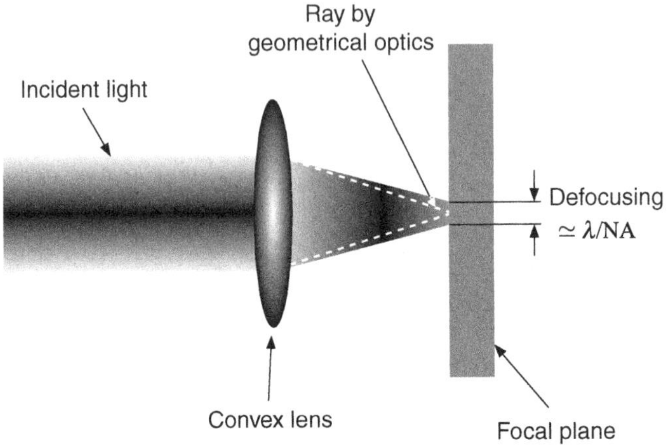

Fig. 1.7. Schematic representation of residual defocusing

A laser beam also diverges due to diffraction. Its divergence angle is expressed as $\lambda/\pi w$ (radian), where w is the spot size of the laser beam [1.3]. For example, the divergence angle is about 0.3 mrad for $\lambda = 1\,\mu\text{m}$ and $w = 1\,\text{mm}$, which means that the 1 mm spot size of the laser beam becomes as large as 3 mm after propagating 10 m. Although this size is much smaller than that of solar light, flash lamps, and light bulbs, it gives undesirable effects in photonic systems.

In the case of optical disk memory, storage and readout of pits smaller than λ/NA are not possible. Shorter wavelength lasers have been intensively developed in order to decrease the diffraction-limited pit size, i.e., the major efforts shown in Fig. 1.2 for increasing storage density have aimed to use shorter wavelength light for storage and readout. However, the upper limit of the storage density achieved using visible light is several $10\,\text{Gb/in}^2$, while the value required in the year 2010 is estimated to be more than ten times greater.

Semiconductor lasers, optical waveguides, and optical switching devices have to confine the light within them for effective operations. In the case of a semiconductor laser, its active layer has to be larger than the diffraction-limited volume, i.e., λ^3, for this confinement. In the case of an optical fiber, its core diameter has to be larger than λ. These examples mean that the sizes of photonic devices cannot be smaller than the wavelength of light, which is the diffraction-limited size of the photonic device. However, sizes of photonic switching devices for optical fiber communication systems in the year 2015 must become smaller than the diffraction-limited size.

The narrowest linewidth of the pattern fabricated by photolithography is also limited by diffraction. The progress in decreasing the pattern size shown in Fig. 1.5 has been the result of efforts to use shorter wavelength light to decrease the diffraction-limited value. However, further shortening of the wavelength requires gigantic and expensive light sources which can become prohibitive when developing practical microfabrication systems. For visible light sources, the 30–70 nm linewidth for 64–256 Gb DRAMs is far beyond the diffraction limit.

To summarize, miniaturization of optical technology is not possible as long as conventional propagating light is used. This is the deadlock imposed by diffraction of light. One must go beyond the diffraction limit in order to open up a new field of optical technology. This field is called nanophotonics.

Problems

Problem 1.1

In order to estimate the value of the divergence angle due to diffraction, Fig. 1.8 shows a slit on a plane Σ at $z = 0$, where the width of the slit is a. The plane light wave of wavelength λ emitted from the light source S is incident on Σ. It can be considered that a point source is generated at the point $P(x = x_1)$ by this incident light. A spherical lightwave is emitted from this source and propagates into the space $z > 0$. In this context, solve the following two problems.

(a) Derive the light intensity at the point Q $(x = x_2)$ in the plane Σ' at $z = z$. Assume that the angle ψ between the z-axis and segment PQ is negligibly small, and that the distance r between P and Q is much longer than λ.
(b) Derive the divergence angle of the light after passing through the slit.

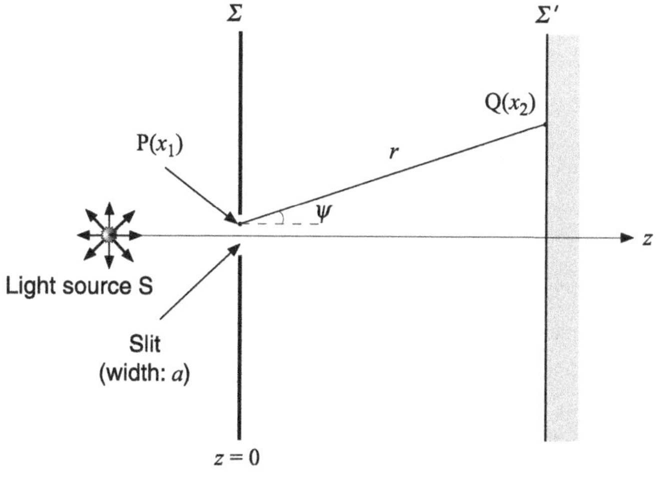

Fig. 1.8. Schematic representation of the diffraction of light by a slit

Problem 1.2

Replace the convex lens in Fig. 1.7 by a cylindrical convex lens with width a and focal length f. By illuminating this lens with light with wavelength λ, derive its spot size on the focal plane.

2 Breaking Through the Diffraction Limit by Optical Near Field

The present chapter describes how to realize nanophotonics for the requirements of the 21st century. This is possible by generating and utilizing an optical near field. Sections 2.1 and 2.2 describe the principles for generating and detecting the optical near field. These reviews introduce the theoretical discussions in Chaps. 4–8.

2.1 Generation of Optical Near Field

How can nanophotonics meet the requirements of the 21st century as presented in Chap. 1? Since light exhibits both particle and wave properties (see Sect. 1.1), one may propose to use the particle property rather than the wave property of light. However, this would not appear to be a good answer, because the particle property of light reviewed in Sect. 1.1 does not mean that light can be confined in a limited space. It means that the energy of light can take a discrete value, as is the case for confined electrons. That is, conventionally used light propagates and spreads in space, i.e., it is not a spatially localized particle. Is it then possible to generate spatially localized light? The main topic discussed in this book seeks to answer this question. And the answer is affirmative. Such light is called an optical near field.

Figure 2.1 shows how to generate an optical near field on a small sphere S with conventionally propagating incident light. Here, it is assumed that the radius a of the sphere S is much smaller than the wavelength λ of the incident light. The scattered light 1 in this figure represents the incident light scattered by the sphere S. However, it should be noted that an optical thin film with thickness about a is also generated on the surface of the sphere S. This is called an optical near field.[1] [2] Since $a \ll \lambda$, the volume of this optical near field is much smaller than the diffraction-limited value. However, it cannot be separated from the sphere S because it is localized on S.

[1] The term 'optical thin film' is used only to give an intuitive image of the nature of the optical near field. Chapter 9 presents a more appropriate term, i.e., an optical cloud localized on the material.

[2] The frequency of the optical near field is equal to that of the incident propagating light.

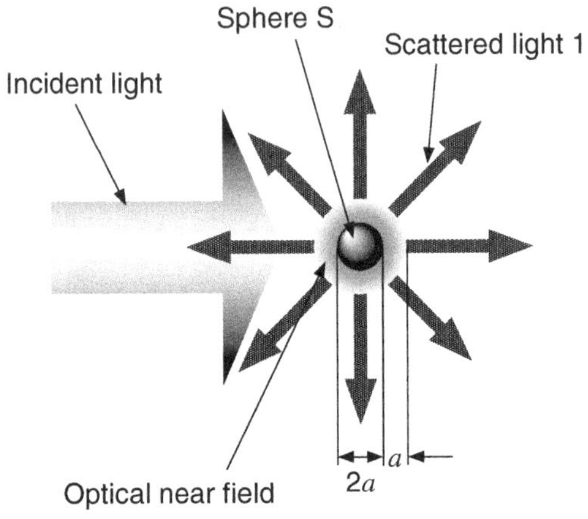

Fig. 2.1. Schematic representation of an optical near field

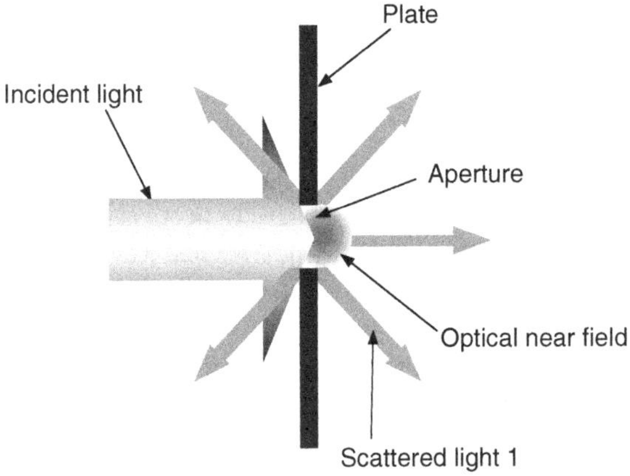

Fig. 2.2. Schematic representation of an optical near field generated by a small aperture

As an alternative case, Fig. 2.2 shows that the optical near field is generated on a circular aperture with a sub-wavelength diameter. With conventionally propagating incident light, a hemispherical optical near field is generated simultaneously with the scattered light 1 on the aperture, and its diameter is close to that of the aperture. In summary, Figs. 2.1 and 2.2 show that the optical near field and scattered light 1 are generated by a sphere S or an aperture. The scattered light propagates to the far field, and exhibits the properties of a light wave as reviewed earlier in Sect. 1.3. In the following, we will not consider the light propagating in the far field but focus on the optical near field whose size depends on the size of the sphere S or the aperture.

Why is the thickness of the optical near field about a? This is explained using Fig. 2.1. If the sphere S is assumed to be made of a dielectric such as

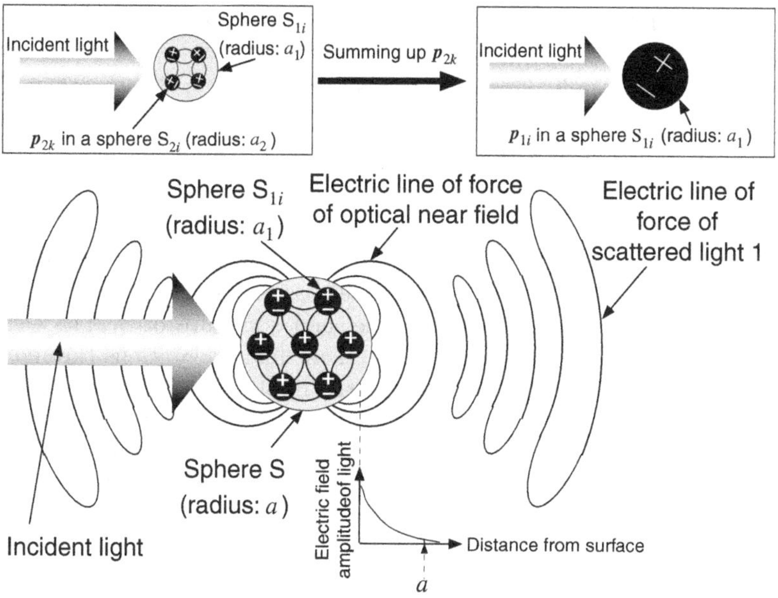

Fig. 2.3. Detailed explanation of the generation of an optical near field

a glass, electrons stay around the nuclei in the atoms which compose S.[3] By shining the light on S, nuclei and electrons in atoms of S are displaced from their equilibrium positions due to the Coulomb forces from the electric field of incident light. Their displacement directions are opposite because the nuclei and electrons are oppositely charged. Thus, pairs of oppositely displaced charges, i.e., electric dipoles are generated. Further, the vector representing the product of electric charges and the displacement vector of this electric dipole is called an electric dipole moment.[4]

The electric dipole moments in S are mutually attracted or repelled due to the Coulomb forces between the oppositely charged nuclei and electrons.[5] Figure 2.3 represents S by an ensemble of smaller spheres with radius $a_1(\ll a)$, where the ith sphere is denoted S_{1i}. In S_{1i}, a great number of electric dipole moments $\boldsymbol{p}_{2k}$ (the kth electric dipole moment in S_{1i}) arrange their orientations as a result of the attractive and repulsive Coulomb interactions with the surrounding electric dipoles. As a result of this arrangement, the vectorial sum of $\boldsymbol{p}_{2k}$ produces a large electric dipole moment $\boldsymbol{p}_{1i}$ in S_{1i}. Figure 2.3 represents the generated $\boldsymbol{p}_{1i}$ in S_{1i}. It should be noted that the

---

[3] The discussion in this section is effective even if S is made of a metal or semiconductor.

[4] Refer to (A.25) of Appendix A for the definition of the electric dipole moment.

[5] It should be noted that the electric field is equivalent to the magnitude and direction of the Coulomb force.

size of S_{1i} has to be small enough for S_{1i} to be considered as having a single electric dipole moment $\boldsymbol{p}_{1i}$.

These electric dipole moments $\boldsymbol{p}_{1i}$ also attract or repel each other. Here it should be noted that they oscillate with the oscillating electric field of the incident light. Figure 2.3 is a schematic representation of such oscillating electric dipole moments, showing how the orientation of $\boldsymbol{p}_{1i}$ is governed by the following two factors:

1. The first factor is the direction of the electric field vector of the incident light. This orientation can be considered as constant in S because $a \ll \lambda$.
2. The second factor is the shape, size, and structure of S. The electric dipole moments $\boldsymbol{p}_{1i}$ in S_{1i} determine their final orientations through the Coulomb interaction with other electric dipole moments $\boldsymbol{p}_{1i}$ in order to maintain the shape, size, and structure of the sphere S. Since the orientation of the electric field vector of the incident light is constant in S, as pointed out in (1), the orientations of $\boldsymbol{p}_{1i}$ are independent of the wavelength and phase of the incident light. They thus depend only on the shape, size, and structure of S.

Figure 2.3 also shows the electric lines of force representing the magnitudes and orientations of the Coulomb forces. These lines connect the electric dipole moments $\boldsymbol{p}_{1i}$. Since the magnitude and direction of the Coulomb force is equivalent to the electric field as pointed out above, these electric lines of force show that $\boldsymbol{p}_{1i}$ generates the electric field. An important feature is that these electric lines of force are found, not only in S, but also on the surface of S. The electric field represented by the electric lines of force on the surface of S corresponds to the optical near field.

These electric lines of force of the optical near field emanate from one electric dipole moment and terminate at another. They tend to take the shortest possible trajectory. As a result, most of them are located in close proximity to the surface of S. This is the reason why the optical near field is very thin. Although we shall later explain quantitatively why the thickness is about a,[6] a qualitative understanding can be obtained by the following considerations. The thickness is independent of the wavelength of the incident light because the two above-mentioned factors determining the orientations of $\boldsymbol{p}_{1i}$ are independent of the wavelength. On the other hand, the only spatial quantity included in the two factors is a, which means that the quantity governing the thickness can only be a.

More quantitatively, the spatial distribution of the optical near field energy decreases rapidly as we move away from the surface, as shown by the small graph in Fig. 2.3. Its energy becomes negligibly small at distance a from the surface of S. This is caused by the fact that most of the electric lines of force are located in close proximity to the surface.

[6] See Sects. 4.3.2 and 8.3

As mentioned above, Fig. 2.3 shows a schematic representation of the electric dipole moments $\boldsymbol{p}_{1i}$, which oscillate synchronously with the oscillating electric field of the incident light. Two kinds of electric field are generated from such oscillating electric dipole moments. One is the optical near field whose electric lines of force come out of one electric dipole moment and terminate at another. The other is the electric field whose electric lines of force form a closed loop, which propagates to the far field. Figure 2.3 also shows these closed loops, which represent the scattered light 1 shown in Fig. 2.1.[7]

Before closing this section, note that the radius a_i of the sphere S_{1i} in S is optional. It can be fixed in such a way that S_{1i} is sufficiently small to be represented by a single electric dipole moment $\boldsymbol{p}_{1i}$. When the observation is made in a region closer to S_{1i}, the latter must be divided into smaller spheres S_{2i} with radius a_2 ($a_2 \ll a_1$), depending on how close the observation point is to the surface of S_{1i}. In this case, the thickness of the optical near field observed on the surface of S_{1i} is about a_1. Further, if the optical near field is observed in close proximity to S_{2i}, it must be divided into smaller spheres S_{3i}. By dividing the sphere into smaller ones depending on the size and position of observation, the discussion regarding Fig. 2.3 can be repeated. This means that there exists a hierarchy within the theoretical model, depending on the size of optical properties observed.

However, this hierarchy is not infinite. If the sphere is repeatedly divided, it will eventually exhibit specific characteristics depending on its size, which are different from those of a bulk sphere. For example, the optical and electrical properties of a nanometric glass differ from those of a bulk glass. The behavior of an ensemble of atoms in a nanometric material becomes very different from that in a bulk system, and this again is different from the behavior of a single atom. The range of sizes between a bulk material and a single atom is called mesoscopic. This is one of the key subjects in material science and technology today.

This finite hierarchy is a common concept in modern science, suggesting that there is an optimum theoretical model ranging from classical macroscopic to quantum atomic theories depending on the size of the physical quantities under study. This concept was also proposed by Democritus, an ancient Greek philosopher. He assumed that there exists a smallest material which survives when a bulk material is repeatedly divided. He called this the atom. On the basis of the modern quantum theory of the atom, it can be further divided into a nucleus and electrons. However, their optical and electrical properties are different from those of the atom, i.e., the specific properties of atoms

[7] Figure A.3 in Appendix A shows the electric lines of force generated from a single oscillating electric dipole moment. This figure also shows two kinds of line. One comes out of the top of the electric dipole moment and terminates at the tail. This corresponds to the optical near field. The other, represented by a closed loop, propagates to the far field. This corresponds to the scattered light. Figure 2.3 shows these lines generated from multiple electric dipole moments.

are lost by this division. An important feature is found by considering the opposite approach, i.e., if one tries to study the properties of the atom by evaluating the characteristics of a mesoscopic material, one need not consider the behavior of the nuclei and electrons in this material. This is the modern concept of the atom.

Supplement: Total Reflection and Evanescent Light

First, assume that there is a planar boundary between materials 1 and 2 with refractive indices n_{h} and n_{l}, respectively ($n_{\mathrm{h}} > n_{\mathrm{l}}$). When light with incidence angle θ_1 is propagated from material 1 to the boundary, part of the power is reflected back whilst the rest of the power is transmitted to material 2 with angle of refraction θ_2. The relation between these angles is given by Snell's law, which is expressed as

$$n_{\mathrm{h}} \sin \theta_1 = n_{\mathrm{l}} \sin \theta_2 \ . \tag{2.1}$$

It is found that $\theta_2 > \theta_1$ because $n_{\mathrm{h}} > n_{\mathrm{l}}$. The maximum of θ_1 is derived by substituting angle $\theta_2 = 90°$ into this equation. It is called the critical angle θ_{c} and is given by

$$\sin \theta_{\mathrm{c}} = \frac{n_{\mathrm{l}}}{n_{\mathrm{h}}} \ . \tag{2.2}$$

There is no transmission if $\theta_1 > \theta_{\mathrm{c}}$, in which case we have total reflection. Figure 2.4 shows this situation. In the case when materials 1 and 2 are water and air, the value of θ_{c} is 48.8° because n_{l} and n_{h} are 1.0 and 1.33, respectively.

Next, assume that the incidence angle is larger than the critical angle, i.e., $\theta_{\mathrm{c}} < \theta_1 < 90°$. Although a real refraction angle θ_2 does not exist, the

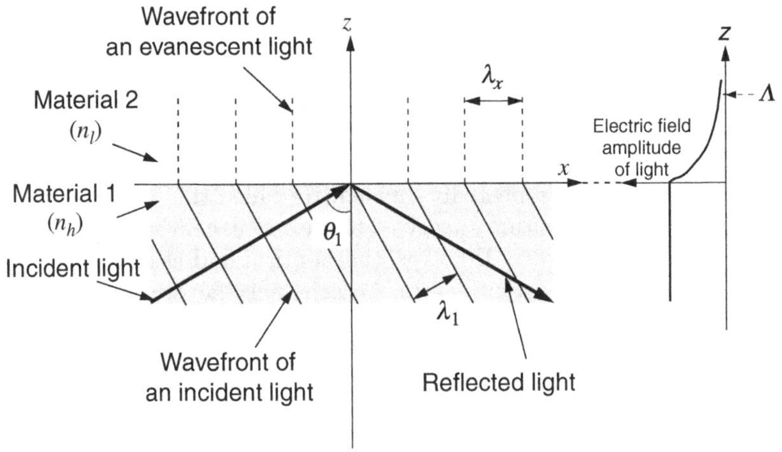

Fig. 2.4. Schematic representation of evanescent light under the condition of total reflection

mathematical expression for θ_2 can be obtained by substituting (2.1) into the relation $\sin^2\theta_2 + \cos^2\theta_2 = 1$ to give

$$\cos\theta_2 = \pm \mathrm{i}\sqrt{\frac{1}{n^2}\sin^2\theta_1 - 1}\ , \tag{2.3}$$

where n is defined as $n_\mathrm{l}/n_\mathrm{h}$ and is called the relative refractive index. If it is assumed that some light penetrates into material 2 in Fig. 2.4, its electric field is given by

$$E(x,z) = T_0 \exp\left[-\mathrm{i}\omega t + \mathrm{i}k_2(x\sin\theta_2 + z\cos\theta_2)\right]\ , \tag{2.4}$$

where $\omega(= 2\pi\nu)$ is an angular frequency. The wave number k_2 in material 2 is given by $k_2 = 2\pi/\lambda_2$, where λ_2 is the wavelength in material 2. Inserting (2.1) and (2.3) into this equation, one obtains

$$E(x,z) = T_0 \exp\left[-\mathrm{i}\omega t + \mathrm{i}k_2\left(\frac{x}{n}\sin\theta_1 \pm \mathrm{i}z\sqrt{\frac{1}{n^2}\sin^2\theta_1 - 1}\right)\right]\ . \tag{2.5}$$

The negative sign in $\pm$ in this equation has to be neglected in order to keep the value of $|E(x,z)|$ finite as $z \to \infty$. As a result, $E(x,z)$ is given by

$$E(x,z) = T_0 \exp\left(-\mathrm{i}\omega t + \mathrm{i}k_2\frac{x}{n}\sin\theta_1\right)\exp\left(-k_2 z\sqrt{\frac{1}{n^2}\sin^2\theta_1 - 1}\right)\ . \tag{2.6}$$

This equation shows that $|E(x,z)|$ decreases exponentially with increasing z, and it is e^{-1} times the value of $|E(x,0)|$ at $z = 1/k_2\sqrt{(1/n)^2\sin^2\theta_1 - 1}$. Thus, the decay length is defined by

$$\Lambda \equiv \frac{1}{k_2\sqrt{(1/n^2)\sin^2\theta_1 - 1}}\ , \tag{2.7}$$

whose value is of the same order as λ_2. Equation (2.6) shows that a surface wave exists on the boundary of material 1 and 2. This is called evanescent light. Further, the first exponential term in this equation shows that this light propagates along the x-axis with wavelength

$$\lambda_x = \frac{n\lambda_2}{\sin\theta_1} = \frac{\lambda_1}{\sin\theta_1}\ , \tag{2.8}$$

where λ_1 is the wavelength in material 1.

The evanescent light does not carry energy away from the planar surface (along the z-axis) because it propagates on the surface (along the x-axis). This feature is analogous for the optical near field shown in Fig. 2.1. However, the thickness of the optical near field is about the radius of the sphere, while that of the evanescent light is Λ. What is the origin of this difference? Figure 2.5

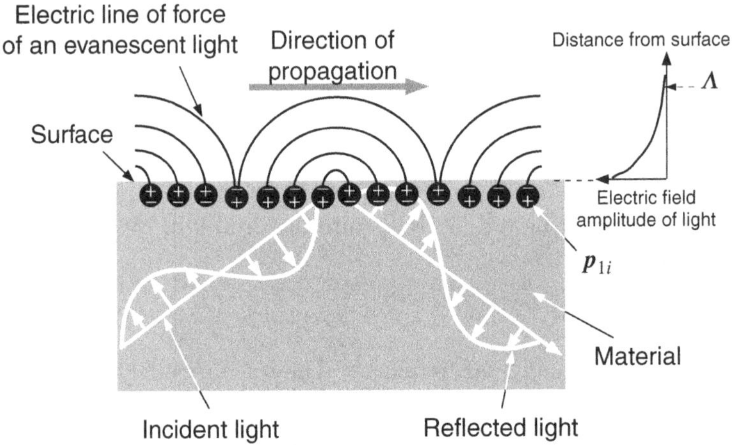

Fig. 2.5. Detailed explanation for the generation of evanescent light

provides the answer to this question. It is drawn up in a similar way to Fig. 2.3. In this figure, the material is assumed to be an insulator such as a glass, and the upper part is assumed to be the vacuum for simplicity. This figure also illustrates the incident and totally reflected light, electric dipole moments, and electric lines of forces gencrated on the boundary. The electromagnetic field represented by these electric lines of force is the evanescent light.

Since the boundary is an infinite plane, the orientations of the electric dipole moments $\boldsymbol{p}_{1i}$ generated in the vicinity of the boundary depend on the spatial phase and wavelength of the incident and reflected light. They are thus arranged periodically. Due to this periodicity, no propagating light is generated on the boundary. In other words, these periodically arranged electric dipole moments do not generate any closed-loop electric lines of force on the boundary, even though they are oscillating with the optical frequency. This is the origin of the total reflection.

However, if there are electric lines of force connecting electric dipole moments on the boundary, which represents evanescent light, the evanescent light then covers the boundary as an optical thin film whose thickness can be calculated by summing all the electric lines of force generated on the surface. The calculated result is given by (2.7), which is of the order of the optical wavelength. On the other hand, in the case of the sub-wavelength-sized sphere S in Figs. 2.1 and 2.2, the orientations of the electric dipole moments $\boldsymbol{p}_{1i}$ in the smaller sphere S_{1i} do not depend on the phase and wavelength of the incident light. It then generates the scattered light 1 and an optical near field as thin as the radius a. It should be pointed out that the planar boundary case shown in Figs. 2.4 and 2.5 is a special case of Figs. 2.1 and 2.2.

The nature of the evanescent light remains within the framework of conventional optics, because the thickness of the evanescent light on a planar surface is of the order of the wavelength of the incident light, which originates from the fact that the spatial distribution of the induced electric dipole moments depends on the wavelength of the incident light. Thus, the use of evanescent light cannot realize any measurement, fabrication, and so on beyond the diffraction limit, and cannot meet the requirements of our society in the 21st century. One has to draw a sharp dividing line between the optical near field on a nanometer-sized surface and the evanescent light on a planar surface.

The physical origin of the optical near field is the same as that of evanescent light. The difference lies in whether the orientation of the induced electric dipole moments depends on the phase and wavelength of the incident light. Finally, it should be pointed out that the term 'optical near field' is more general than the term 'evanescent light' when expressing the nature of the surface electromagnetic field.

2.2 Detection of Optical Near Field

The scattered light 1 in Fig. 2.3 can be detected by a photodetector placed in the far-field region of the sphere S because it is conventional propagating light. However, the optical near field cannot be detected because it is localized on the surface of S and does not carry energy to the far field. In order to detect the optical near field, the method shown in Fig. 2.6a is employed, i.e., the optical near field is disturbed by a secondary sphere P. The disturbed optical near field is converted to propagating light, called scattered light 2, and its energy is transferred to the photodetector to be detected.[8]

Figure 2.6b explains the principle of this detection with the help of electric dipole moments $\boldsymbol{p}_{1i}$ induced in the spheres S_{1i} and electric lines of force. By placing P in the optical near field on the surface of S, some electric lines of force are directed to the surface of P and induce electric dipole moments $\boldsymbol{p}_{1i}$ in P. These electric dipole moments generate the scattered light 2 as well as the optical near field on the surface of P. Measurement of the power of this scattered light 2 corresponds to detection of the optical near field on the surface of S.

The detection of the optical near field described above uses the demolition method, i.e., orientations and positions of the electric dipole moments $\boldsymbol{p}_{1i}$ in S and electric lines of force are varied by the approach of P.[9] Since electric lines of force are newly generated from P, one must consider the spatial distribution of the total electric lines of force on the closely spaced spheres S and P. This means that the two spheres S and P establish a mutually

[8] The rate of energy flow per unit time is called the power, which is the flow rate of the number of photons.

[9] Note that orientations of some electric dipole moments $\boldsymbol{p}_{1i}$ of S in Fig. 2.6b are illustrated as different from those in Fig. 2.3.

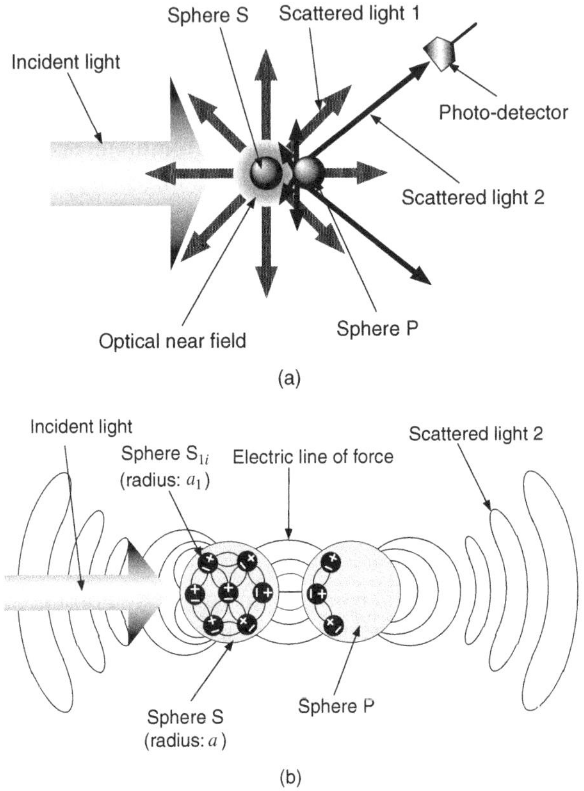

Fig. 2.6. Schematic representation of the detection of an optical near field. (**a**) Basic explanation. (**b**) Detailed explanation

dependent situation to be connected by the electric lines of force. However, this situation becomes valid only in the presence of the incident light. If the incident light is off, they are independent of each other, even though they are very close. Thus, the detection of the optical near field forms a special coupling between the two spheres. It means that this coupling exhibits unique response characteristics to the incident light, which differ considerably from those exhibited by two independent spheres. Hence, the two spheres are coupled to each other in response to the incident light, even though they are mutually isolated material systems. Such an optically coupled condition is called a mesoscopic condition, as pointed out at the end of the last section.

Although the scattered light 2 is generated by disturbing the optical near field on S, one should note that there still exists scattered light 1 which originates from S. Therefore, the scattered light 2 must be selectively detected for sensitive demolition detection of the optical near field on S. For this selective detection, Fig. 2.7 shows the use of a screen to prevent detection of scattered light 1. As a successful device possessing this screening function,

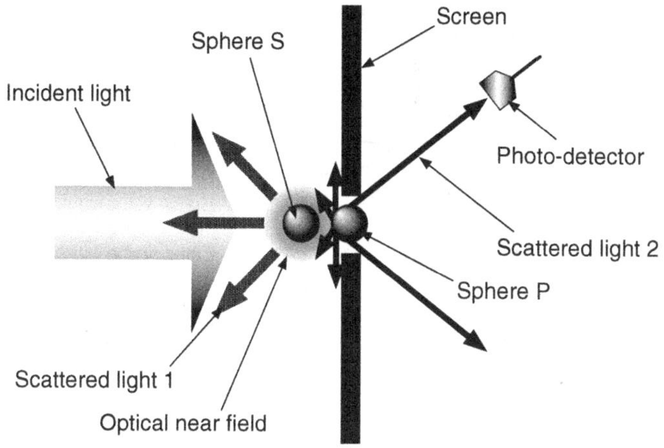

Fig. 2.7. Using a screen to prevent detection of scattered light 1

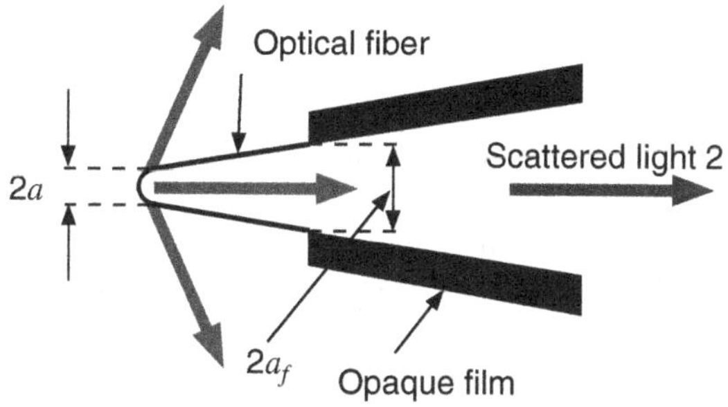

Fig. 2.8. Schematic illustration of a fiber probe (collection mode)

a sharpened optical glass fiber has been used instead of P. This is called a fiber probe, as shown in Fig. 2.8, whose tip radius a corresponds to the radius of P. By inserting this fiber probe into the optical near field on S, the probe tip disturbs the optical near field to generate the scattered light 2. Part of the scattered light 2 is coupled to the fiber probe and transmitted to its end. A photodetector installed at the end measures the transmitted power of the scattered light 2. On the other hand, an opaque film coated on the tapered part and foot of the fiber probe works as a screen to prevent scattered light 1 from coupling to the fiber probe. This is because the foot radius a_{f} of the protruding sharpened core is smaller than the wavelength of the incident light. By this screening, the scattered light 2 is selectively detected. A metallic film of Al or Au has often been used.

By applying this detection method, a novel optical microscope can be made, which is called a near-field optical microscope, and explained schematically in Fig. 2.9. The power of scattered light 2 is measured by inserting a fiber probe into the optical near field on the surface of S. After this measure-

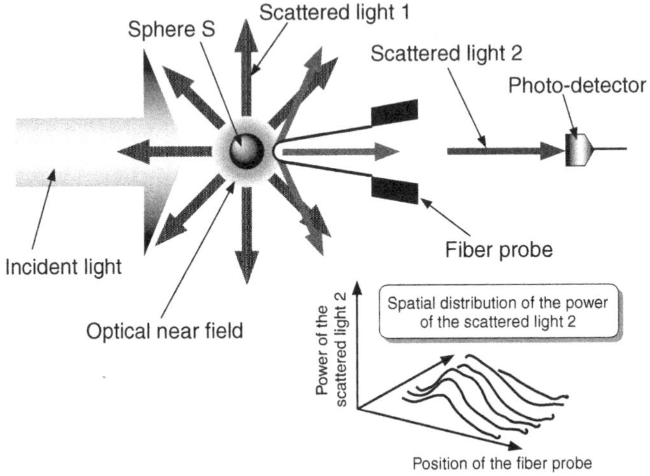

Fig. 2.9. Schematic diagram of a near-field optical microscope (collection mode)

ment, the fiber probe is scanned to another place in the optical near field to measure the power of the scattered light 2 again. By repeating this scanning and power measurement, the spatial distribution of the power of scattered light 2 is obtained as a function of the position of the fiber probe. Since scattered light 2 originates from the optical near field, this distribution represents the spatial distribution of the optical near field energy, and also the shape of S. As a result, this distribution represents a microscopic image of S.[10]

How high is the resolution of this microscope? It depends on the tip size a of the fiber probe, i.e., the radius of the sphere P, because it is determined by the fractional volume of the optical near field from which the scattered light 2 is generated. Thus, the smaller the tip size, the higher the resolution. Since the discussion of resolution given above is independent of the wavelength of the incident light, magnifications much higher than the diffraction-limited value can be realized by using a sub-wavelength-sized tip.

Figures 2.10a and b illustrate the last comment of the present chapter. Comparing Fig. 2.10a with 2.6a, the positions of the light source and photodetector are exchanged. It would be more appropriate to say that the roles of the spheres S and P are exchanged. That is, Fig. 2.10a represents light propagating from the light source to the sphere P and the sphere S is then illuminated by the optical near field on the surface of P. This optical near field is disturbed by S and the converted scattered light 2 propagates to the photodetector.

Figure 2.10b shows the sphere P of Fig. 2.10a replaced by a fiber probe, where the light from the source is injected into the end of the fiber probe to

[10] The sphere S stands for the sample under microscopic measurement, whilst sphere P is named after the probe.

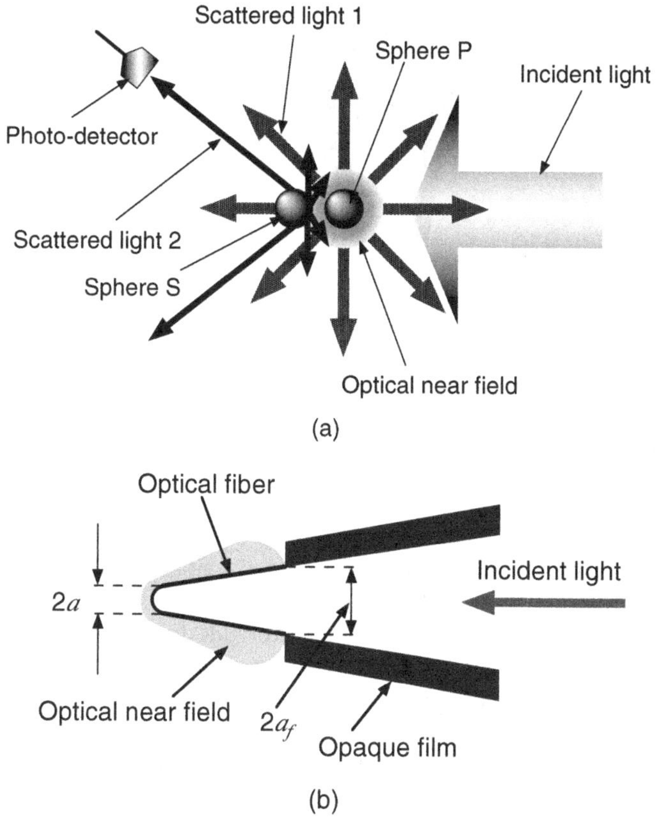

Fig. 2.10. Illumination mode of operation. (**a**) Principle. (**b**) How to use a fiber probe

generate the optical near field on the tip. The scattered light 2 is generated in S by the optical near field, and its power is measured by the photodetector while scanning the fiber probe. The spatial distribution of the measured power as a function of the position of the fiber probe gives the microscopic image of S. This mode of operation is called illumination mode, wherein the fiber probe is used as light source for the optical near field to illuminate the sample. The mode of operation shown in Figs. 2.6a, 2.8, and 2.9 is called collection mode, because the optical near field is scattered and collected by the fiber probe.

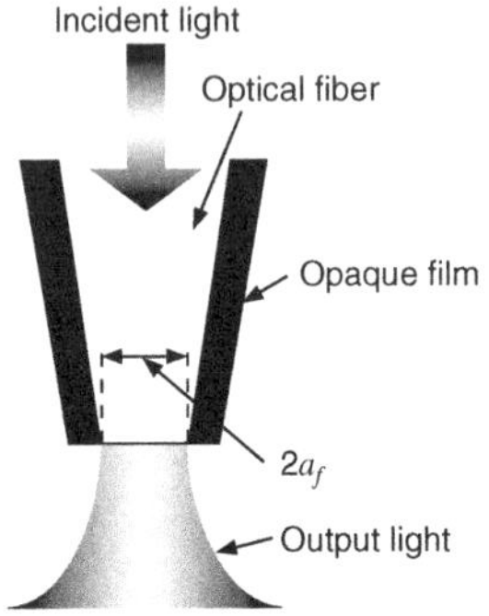

Fig. 2.11. Schematic illustration of a fiber, as found in certain references

Problems

Problem 2.1

Figure 2.11 shows the cross-sectional profile of a fiber probe. It is a scanning electron microscope image which has been shown in certain references. In this figure, the sharpened core does not protrude from the opaque metallic film. The top of this fiber probe is called an aperture, by analogy with the aperture of Fig. 2.2. Its radius corresponds to a_f in Fig. 2.8. It is sometimes claimed that the spatial resolution of a near-field optical microscope obtained using this fiber probe is determined by a_f. Is this claim correct?

Problem 2.2

Characteristics of the optical near field are independent of the wavelength of the incident light. So what color does the optical near field have?

3 Past and Present of Near-Field Optics

The present chapter discusses practical aspects of the novel science and technology that uses optical near fields. The history of research and development is described in Sect. 3.1. Section 3.2 describes the structure and performance of probes, which are key devices for generating or detecting optical near fields. The present status of research and development is reviewed in Sect. 3.3, including applications to microscopy, spectroscopy, fabrication, optical disk memory, and atom manipulation. Finally, Sect. 3.4 overviews possible trends in the novel fields of optical science that will be founded by exploiting optical near fields.

3.1 History and Progress

A primitive proposal to apply the optical near field on a small aperture (see Fig. 2.2) to high resolution optical microscopy was made as early as 1928 [3.1]. However, the theoretical discussion was limited to the framework of wave optics. Consequently, it referred neither to the inherent nature of the optical near field nor to the more promising applications in spectroscopy, fabrication, and manipulation. After this proposal, no significant research and development was carried out for about half a century. Modern studies on the optical near field started in the early 1980s. Early in 1982, the fabrication of a fiber probe was launched in Japan by sharpening an optical fiber [3.2]. Soon afterwards, in 1984 [3.3] and 1986 [3.4], experimental results on near field optical microscopy were published. These were obtained independently of the Japanese work by using the sharp edge of a quartz crystal as a probe. Research and development trends in the 1980s were limited to microscopy, and until today, most of the work has concentrated on applications for scanning probe microscopy to study optical properties of organic/inorganic materials and biological samples.

Since the early 1980s, just after launching the fiber probe fabrication, one of the authors (M.O.) realized that fabrication was a more essential application of the optical near field, and that high resolution fiber probes must be fabricated in a reproducible manner in order to realize these applications. Following this realization, remarkable progress was made in fiber probe fabrication by developing a chemical etching process. Applying these fiber probes to near-field optical microscopy, high resolution images were obtained, proving the superiority of chemically-etched fiber probes over other probes. This is because the chemical etching process has a very high reproducibility as compared with other processes such as heating–pulling methods. Applications to spectroscopy, fabrication, optical disk memory, and atom manipulation were developed using these high performance fiber probes.

3.2 Probe Technology

A probe is a key device for optical near field science and technology. Satisfactory and reliable results cannot be obtained without a high quality probe. The following three basic features are required of the probe:

1. Its tip must be small in order to treat the small optical near field.
2. The conversion efficiency of the optical near field from/to propagating light must be high, for efficient generation and detection of the optical near field.
3. Screening capability must be included to discriminate the optical near field from unwanted propagating light, as shown in Fig. 2.7.

In order to meet these requirements, an optical glass fiber is used to fabricate the probe, called a fiber probe, as shown in Figs. 2.8 and 2.10b. The first requirement above is fulfilled by sharpening the fiber to decrease the tip radius a. The second feature is achieved by controlling the shape of the tapered part between the tip and the foot. As a result of this control, the propagating scattered light 2 generated by disturbing the optical near field with the tip is effectively guided to the foot and finally transmitted to the photodetector in the case of collection mode. In the case of illumination mode, the propagating incident light reaching the foot is effectively guided to the tip in order to generate an optical near field. After this sharpening, the taper is coated by an opaque metallic film to fulfill the third requirement.

Figure 3.1a shows the cross-sectional profiles of an optical glass fiber which is chemically etched in a buffered HF solution in order to fulfill conditions (1) and (2) above. The left-hand picture of Fig. 3.1b shows a SEM image of a sharpened fiber with tip radius $a \leq 2\,\mathrm{nm}$. The right-hand picture of Fig. 3.1b shows a SEM image of the foot radius $a_{\mathrm{f}} = 15\,\mathrm{nm}$ after coating with a metallic film. Figure 3.1c shows cross-sectional profiles of sharpened fibers which have been used for various applications. In this figure, the high-resolution-type fiber probe has a relatively small tip radius a, as is the case in Fig. 3.1b.

A high-efficiency-type fiber probe is fabricated in order to fulfill requirement (2). The conversion efficiency, defined as the ratio of the converted energy of the optical near field to the energy of the incident light, is required to be higher than 10%. T. Yatsui et al. demonstrated a higher conversion efficiency than 10% in illumination mode [3.5]. Figure 3.1d shows the SEM image of such a high-efficiency-type probe, which is referred to as double-tapered-type [3.6]. Double taper can decrease the length of the tapered part, thereby achieving high efficiency. Its tip is buried into the metallic film to make a flat top surface for spectroscopic applications (see Sect. 3.3.2). The radius of the central hole on the metallic top surface, sometimes called an aperture, corresponds to the foot radius a_f of Fig. 2.10b.[1] It should be noted that the aperture radius a_f of the fiber probe does not determine the resolution because the contribution from the tip radius a at the center of the aperture remains.[2]

As a combination of high-resolution-type and high-efficiency-type probes, a triple-tapered fiber probe has been developed [3.7]. Further, a functional-type probe has been developed by fixing light-emitting dye molecules on the tip [3.8]. By selecting optical fiber materials, high efficiency fiber probes have been developed in the visible, infrared, and ultraviolet regions [3.7]. Recently, by applying a microfabrication process to a silicon crystal substrate, a two-dimensional array of probes has been developed for high efficiency and fast scanning (see Fig. 3.15) [3.9]. A probe using a cantilever technology like an atomic force microscope has also been fabricated [3.10] (see the supplement to Sect. 3.3.1).

In order to fulfill the third requirement listed above, the foot of the fiber probe has to be coated with an opaque material, as shown in Figs. 2.8 and 2.10b. For this purpose, a metallic film has been used. This avoids the unnecessary detection of the scattered light 1 of Fig. 2.2 in the fiber (collection mode) and the leakage of the incident light through the tapered part (illumination mode). These are possible because the foot radius a_f in Figs. 2.8 and 2.10b is much smaller than the wavelength of the propagating light, i.e., propagating light such as the scattered light 1 and incident light suffer large transmission losses by such a sub-wavelength aperture.

Instead of using a fiber probe, a metallic needle has sometimes been used. This was originally used as a probe for a scanning tunneling microscope (see the supplement to Sect. 3.3.1). It is called an apertureless probe, which fulfills requirement (1). Further, the metallic probe is capable of highly efficient generation of scattered light 2 by disturbing the optical near field. This is due to the high refractive index of the metal. However, fatal defects are that it does not satisfy requirements (2) and (3).[3]

In order to correct these defects, an advanced fiber probe has been developed, whose SEM image is shown in Fig. 3.1e [3.11]. Requirement (1) is fulfilled because it is made of a sharpened fiber. Further, the scattering effi-

[1] The value of a_f in this figure is 35 nm, i.e., the aperture diameter is 70 nm. However, the value is measured by a SEM. Since the metallic film in the vicinity of the aperture is thin, it leaks light. Therefore, the optically effective radius for generating and detecting the optical near field should be larger than this value.

[2] See Problem 2.1 of Chap. 2

[3] It has an additional defect, as discussed in Problem 3.2 at the end of this chapter.

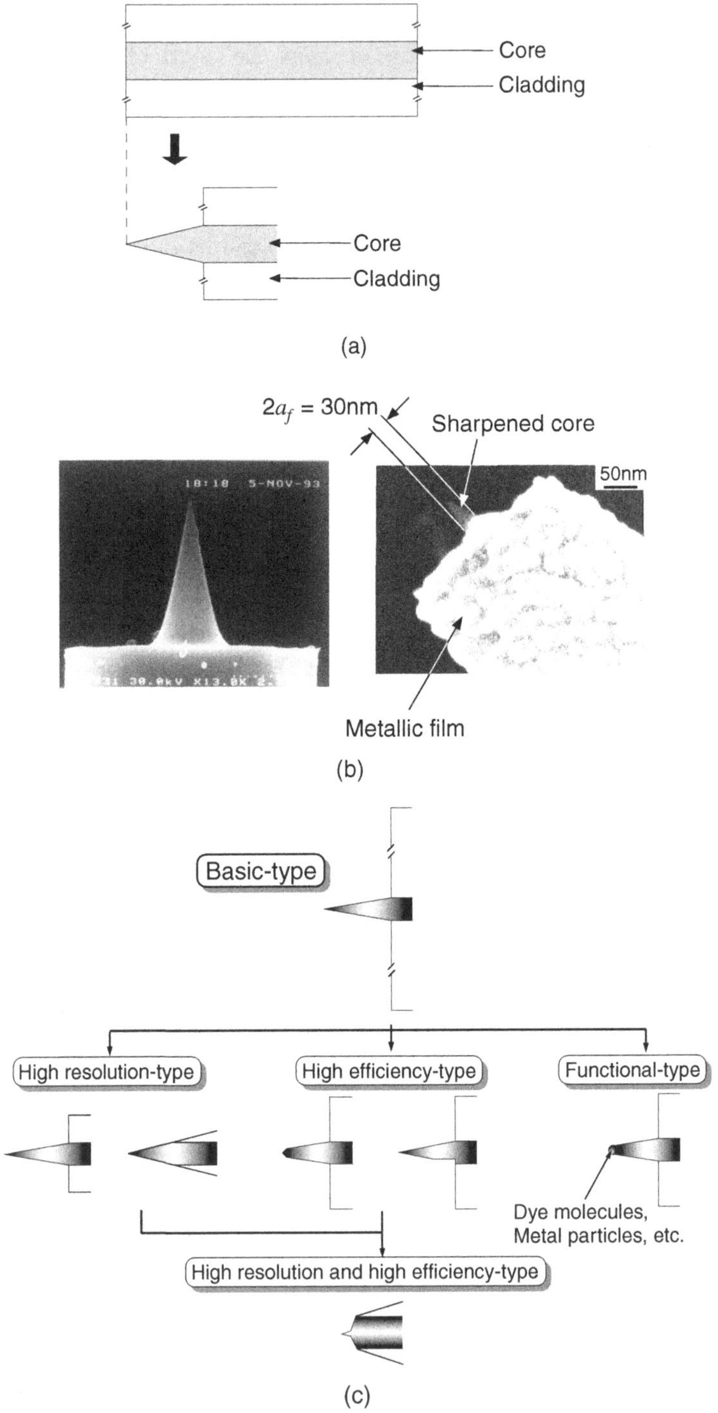
Core
Cladding
Core
Cladding
(a)
$2a_f$ = 30nm
Sharpened core
50nm
18:18 5-NOV-93
Metallic film
(b)
Basic-type
High resolution-type
High efficiency-type
Functional-type
Dye molecules,
Metal particles, etc.
High resolution and high efficiency-type
(c)

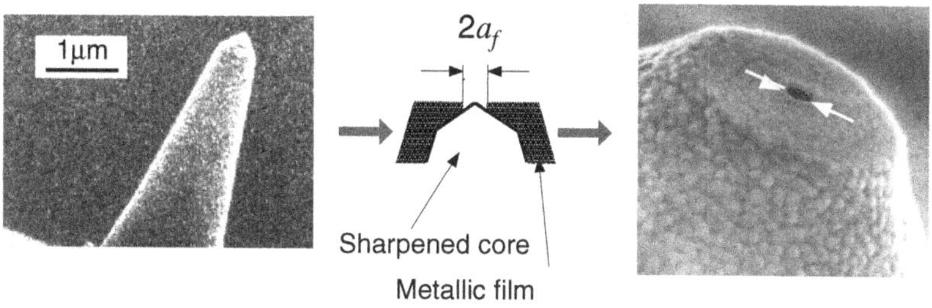

(d)

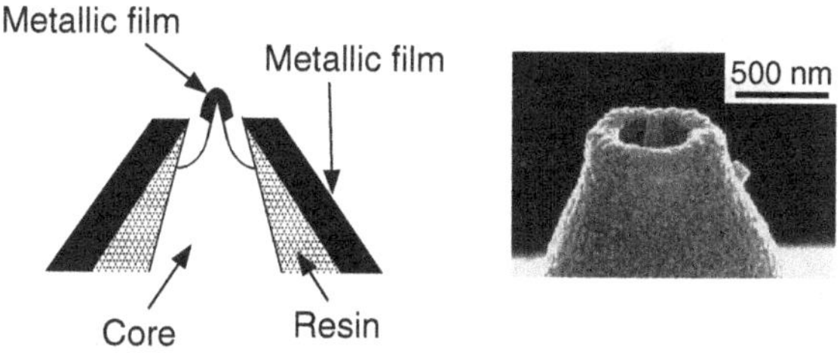

(e)

Fig. 3.1. A fiber probe. (**a**) Method for sharpening an optical glass fiber by chemical etching. (**b**) SEM image of a basic fiber probe. *Left*: Optical glass fiber sharpened by chemical etching. The width of the image is 8.9 μm. *Right*: Metal-coated fiber probe. (**c**) Cross-sectional profiles of sharpened fibers which have been used for various applications of fiber probes. (**d**) SEM image of a double-tapered-type fiber probe. *Left*: Fiber core with a double tapered profile. *Right*: Metal-coated probe tip. *Two white arrows* represent the diameter $2a_f$ (= 70 nm). (**e**) Fiber probe with the metal-coated tip of the sharpened core. *Left*: Schematic explanation of the cross-sectional profile. *Right*: SEM image

ciency is as high as that of the metallic needle because its tip is coated with a metallic film. Requirement (2) is satisfied because its tapered part is controlled by the chemical etching technique. Finally, requirement (3) is fulfilled because its tapered part and foot are coated with an opaque metallic film.

As mentioned above, the profile of a fiber probe can be expressed by a tip radius a, foot radius a_f, and cone angle θ. This profile can be approximately represented by a chain of small spheres which are connected in order of increasing radius, from a up to a_f, as shown in Fig. 3.2a. These spheres disturb the optical near field in the case of collection mode, while they generate the optical near field on their surfaces in the case of the illumination mode. Therefore, in the case of the collection mode, high disturbing efficiency is obtained if the size of the optical near field falls between a and a_f, i.e., the dependence of the efficiency on the size of the optical near field shows the characteristics of a band-pass filter, as shown by the folded lines A and B in Fig. 3.2b. This means that the collection-mode near-field optical microscope

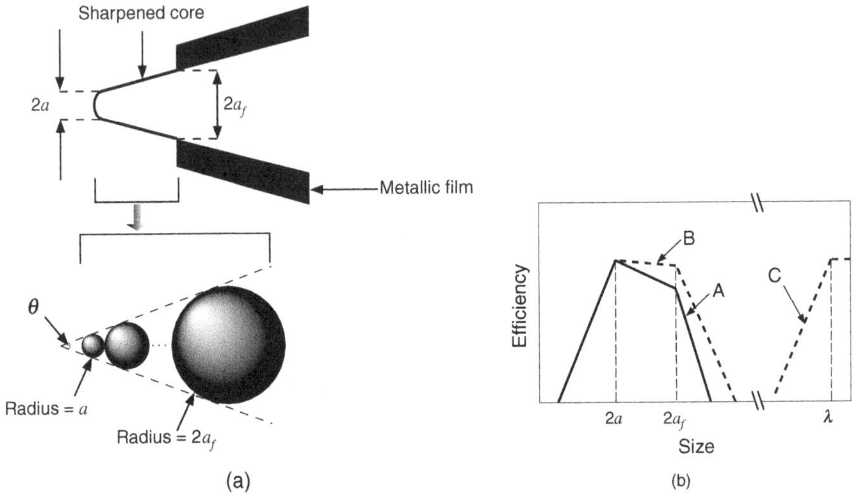

Fig. 3.2. Filtering characteristics of a fiber probe. (**a**) Approximation by a chain of different-sized spheres. (**b**) Relation between the size of the optical near field and the efficiency of generation or detection of the optical near field. Folded lines A and B represent the efficiencies for the fiber probe with smaller and larger cone angles θ, respectively. The folded line C is for a conventional optical microscope

can detect an optical near field whose size falls within the pass band of this spatial filter (i.e., between a and a_{f}). Here, the figure shows that a fiber probe with smaller θ (folded line A) exhibits lower efficiency at a_{f} than that with a larger θ (folded line B), because the sphere of radius a_{f} is farther from the tip in the case of a smaller θ.[4]

This means that the sharper fiber probe can fulfill higher selectivity in picking up the optical near field with size as small as the tip radius a. On the other hand, in the case of illumination mode, the folded lines A and B represent the size dependence of the optical near field energy generated on the fiber probe, i.e., the optical near field with size ranging from a to a_{f} is efficiently generated. These size dependencies of optical near field detection and generation are valid not only for collection and illumination mode near-field optical microscopes, but also for all other applications such as spectroscopy, fabrication, and manipulation.[5]

---

[4] See Sect. 4.3.1

[5] The image formation efficiency of a conventional optical microscope using a combination of lenses shows the characteristics represented by the folded line C in Fig. 3.2b. Due to the diffraction limit of light, it detects light with size larger than the wavelength λ, where $\lambda \gg a, a_{\mathrm{f}}$.

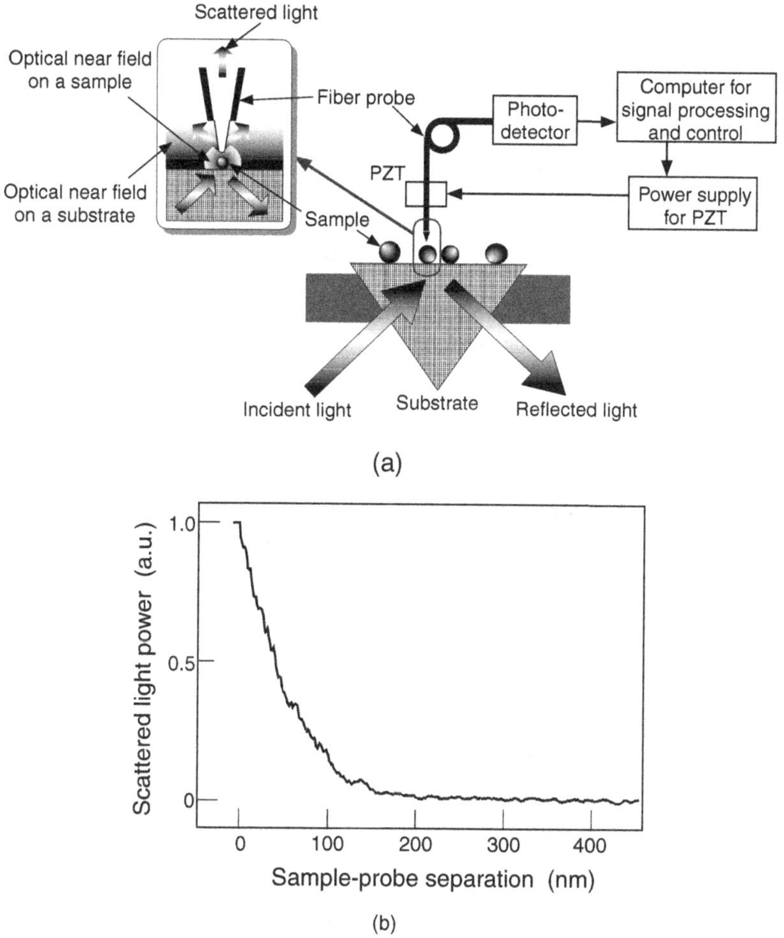

Fig. 3.3. Schematic diagram of the collection-mode near-field optical microscope. (**a**) Configuration of setup. (**b**) Relation between sample–probe separation and power of scattered light 2 measured by disturbing the optical near field on a 30-nm-diameter plastic sphere

3.3 Development of Nanophotonics Using Optical Near Fields

3.3.1 Microscopy

Figure 3.3a is a schematic diagram of a collection-mode near-field optical microscope. In order to generate the evanescent field on the substrate of the sample,its rear surface is irradiated by incident light. The optical near field

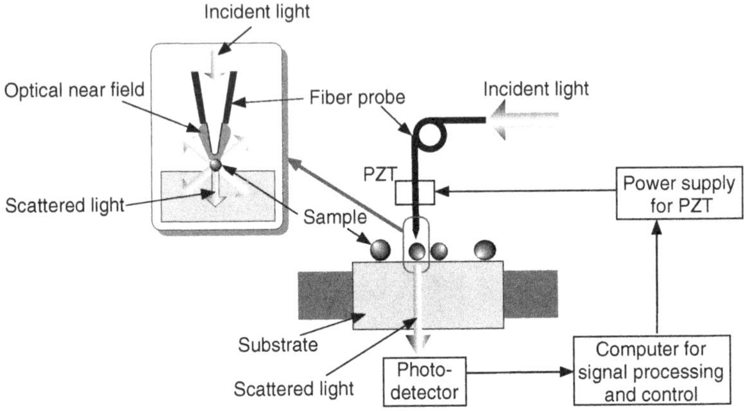

Fig. 3.4. Schematic diagram of the illumination-mode near-field optical microscope

on the sample is generated by this evanescent light.[6] PZT in this figure stands for piezoelectric transducer, which is an electromechanical transducer used to scan a fiber probe precisely. Figure 3.3b represents the relation between the sample–probe separation and the power of scattered light 2 measured by disturbing the optical near field on a 30-nm-diameter plastic sphere whilst vertically scanning the fiber probe. A drastic decrease in detected power is seen with a decay length of about the size of the sphere.

A schematic diagram of an illumination-mode near-field optical microscope is shown in Fig. 3.4. The optical near field is generated on a fiber probe tip which is disturbed by the sample in order to generate the scattered light 2. Although scattered light 2 is detected from the rear surface of the substrate in this figure, it can also be detected from the front surface of the substrate.

The sample–probe separation must be maintained constant in order to scan the probe along the horizontal direction in a stable manner. One popular technique is to detect the magnitude of the mechanical interaction called the shear force between the probe tip and the sample. The vertical position of the probe is feedback-controlled in order to maintain the magnitude of the detected shear force constant. However, it should be noted that the contours of the equipotential surface of the shear force and the equi-energy surface of the optical near field differ from one another, as explained schematically in Fig. 3.5. Thus the probe can cross the equi-energy surface of the optical near field when it is scanned by tracing the equipotential surface of the shear force. In this figure, point A represents such a crossing point. As a result, the power of the detected scattered light 2 varies at this point so that an artifact image of the sample appears. This crossing occurs only because the slope of the equipotential surface of the shear force is sharper than that of

[6] Instead of evanescent light, propagating light can be used to generate the optical near field. However, it is more advantageous to use the evanescent light because one can reduce the power of the scattered light 1 (see Fig. 2.1).

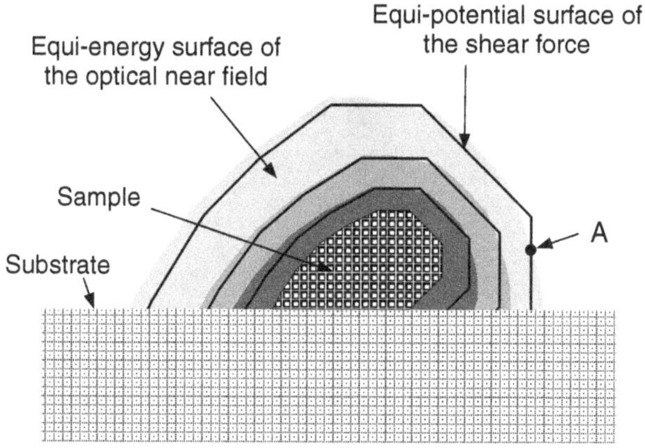

Fig. 3.5. Contours of the equipotential surface of the shear force and equi-energy surface of the optical near field on a sample

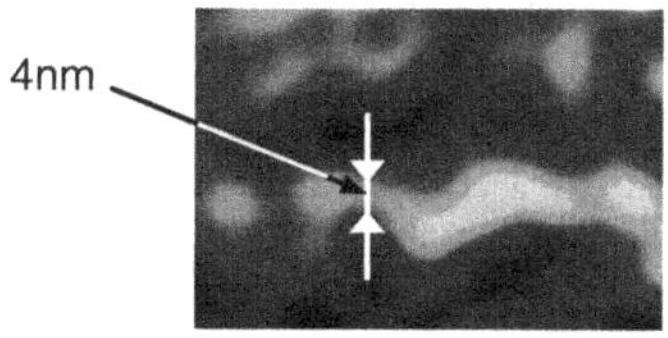

Fig. 3.6. Topographical image of a single string of DNA on an ultrasmooth sapphire surface, taken by a collection-mode near-field optical microscope

the equi-energy surface of the optical near field. This is not related to the high resolution capability of the microscope. The feedback-control system for sample–probe separation should be carefully designed in order to avoid this artifact.

As an example, Fig. 3.6 shows a topographical image of a single string of DNA, i.e., a biological sample, mounted on an ultrasmooth sapphire surface, taken by a collection-mode near-field optical microscope [3.12]. The width of 4 nm in this image exhibits the world record for the highest resolution, which is less than 1/100 of the wavelength of the incident light. This image does not contain any artifact because the sample–probe separation was controlled to maintain the detected near-field optical energy constant, instead of employing the shear force feedback control. The separation was maintained at 1 nm.

Various biological samples can be imaged in the air, as shown in this figure, and imaging in solution has also been demonstrated [3.13]. Imaging in such varied environments has never been possible with an electron microscope.

Supplement: Scanning Probe Microscopes

A family of high resolution microscopes has been developed that is related to the near-field optical microscope. These are the scanning probe micro-

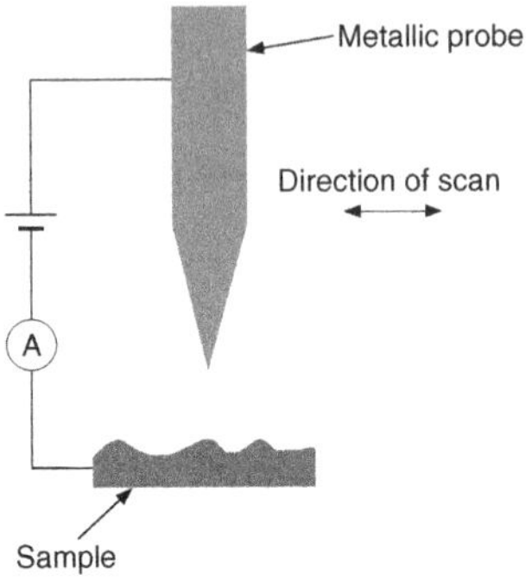

Fig. 3.7. Basic structure of a scanning tunneling microscope (STM)

scopes. Figure 3.7 shows one example. By bringing a metallic probe towards a conductive sample, if it is electrically biased, electrons in the sample tunnel through the potential barrier of the sample surface, thereby producing a tunneling current. A sample image can be obtained by scanning the probe and mapping the spatial distribution of the magnitude of this current. This apparatus is called a scanning tunneling microscope (STM).

In order to explain the principles underlying the generation and detection of the tunneling current, Fig. 3.8a shows the exponentially decaying wave function of an electron which penetrates the potential barrier of the conductive sample surface. By bringing the probe close to the sample surface as shown in Fig. 3.8b, the penetrating wave function can be picked up and converted to a propagating wave function in the probe, whereby a tunneling current is obtained. A topographic image of the sample surface can be obtained with high resolution because of the short penetration length of the wave function. Figure 3.9 shows the experimental result of imaging the two-dimensional array of atoms on a graphite surface. It should be noticed that the spatial layout between the sample and metallic probe is analogous to that of the spheres S and P in Fig. 2.6a. The analogy can also be drawn between the penetrating wave function of the electron and the optical near field of Fig. 2.6b, and between the propagating wave function in the metallic probe and the scattered light 2 of Fig. 2.6b.

Apart from the STM, there are various microscopes which can be categorized as scanning probe microscopes. Among them, the most widely used is the atomic force microscope (AFM) utilizing the atomic force between the sample and probe.[7] AFM has been used to obtain topographical images of

[7] Atomic force is the mechanical force existing between atoms. The potential energy for this force is proportional to R^{-6}, where R is the internuclear distance between atoms. This is derived as follows. Due to the Coulomb interaction between atoms A and B, an electric dipole moment is instantaneously generated in atom A. This electric dipole moment generates an electric field whose magnitude at atom B is proportional to R^{-3} [refer to the last term of (A.28) in Appendix A

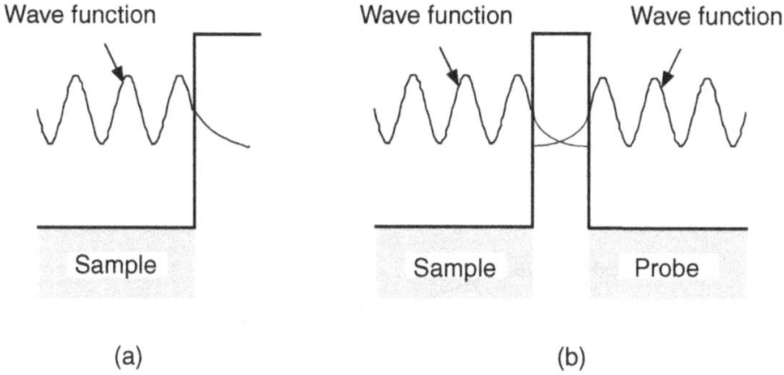

Fig. 3.8. Schematic representation of the principles underlying the scanning tunneling microscope (STM). (**a**) An exponentially decaying electron wave function penetrates the potential barrier of the electrically conductive sample surface. (**b**) The penetrating wave function is picked up by the approaching metallic probe

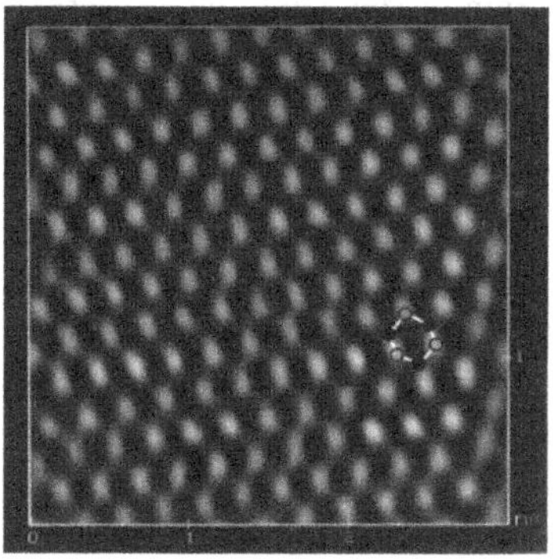

Fig. 3.9. Image of the two-dimensional array of atoms on a graphite surface taken by STM. Image size 3 nm × 3 nm

conducting, insulating, and biological samples in vacuum, air, and solution. The shear force described in Sect. 3.3.1 is one form of the atomic force.

3.3.2 Spectroscopy

The optical near field can be applied, not only for topographical imaging, but also for mapping the spatial distribution of the photon energy emitted from the sample. Further, emission spectra can be measured, i.e., emission

and (D.3) in Appendix D]. By this electric field, atom B generates an electric dipole moment whose magnitude is also proportional to R^{-3}. The magnitude of the electric field at atom A originating from the electric dipole moment in atom B is thus proportional to $R^{-3} \times R^{-3} (= R^{-6})$. This shows that the potential energy is proportional to R^{-6}, and exerts an attractive force towards the ground-state atom, called the van der Waals force.

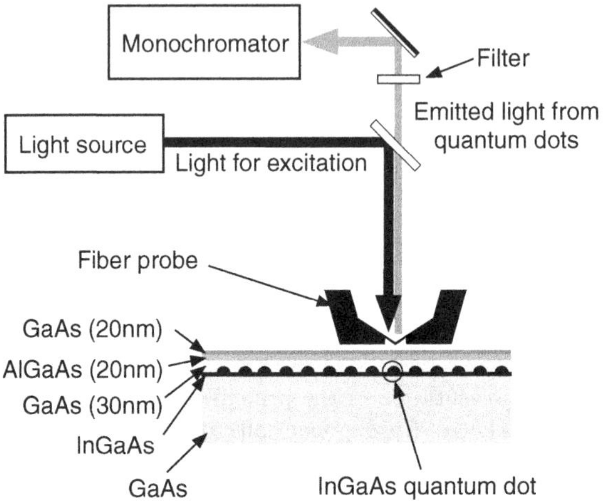

Fig. 3.10. Schematic diagram of the experimental setup to measure the emission spectra of semiconductor quantum dots. The numerical value in brackets represents the thickness of each layer

spectroscopy can be carried out. Figure 3.10 shows an experimental setup to measure the emission spectrum of a nanometric particle of InGaAs called a quantum dot, to be used as an active medium for advanced semiconductor lasers [3.14]. Electronic energy in quantum dots takes discrete values because the size of the quantum dot is as small as the spread of the electronic wave function.[8] The quantum dots shown in this figure have an average diameter of 30 nm and an average height of 15 nm. They are grown on a GaAs substrate with a high density and covered by thin cap layers. Note that topographical images of the quantum dots cannot be obtained from the top surface because of these cap layers.

Free electrons are generated in the cap layers when the sample surface is illuminated by the optical near field on the fiber probe tip. Once these electrons have diffused about 1 μm, they flow into the quantum dots. Light is emitted from the quantum dot by the electronic transition from the conduction band to the valence band. The spatial distribution of the emitted light intensity can be mapped if it is collected by the same fiber probe used to illuminate the sample surface. Emission spectra can also be obtained. Curves A–D in Fig. 3.11 show the emission spectra of quantum dots measured for four different optical excitation power densities at liquid helium temperature. The two peaks connected by broken lines correspond to the two discrete energies of the electrons. Saturation of the peak intensity is observed by increasing the optical excitation power density. Figures 3.12a and b show the spatial distribution of the emission intensity and the ordinal numbers of discrete

[8] This electron corresponds to a quantum mechanical particle in a potential well. Its energy thus assumes discrete values.

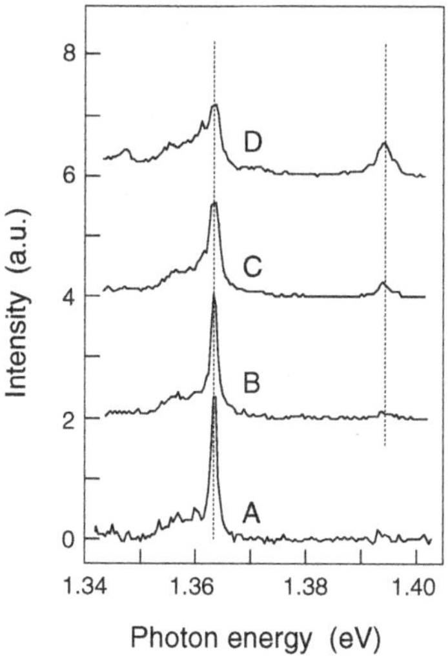

Fig. 3.11. Profiles of emission spectra of InGaAs quantum dots measured at liquid helium temperature. Curves A, B, C, and D represent optical excitation power densities of 1.3 W/cm^2, 5 W/cm^2, 14 W/cm^2, and 28 W/cm^2, respectively. The *two broken lines* connecting the peaks of these curves correspond to discrete electron energies

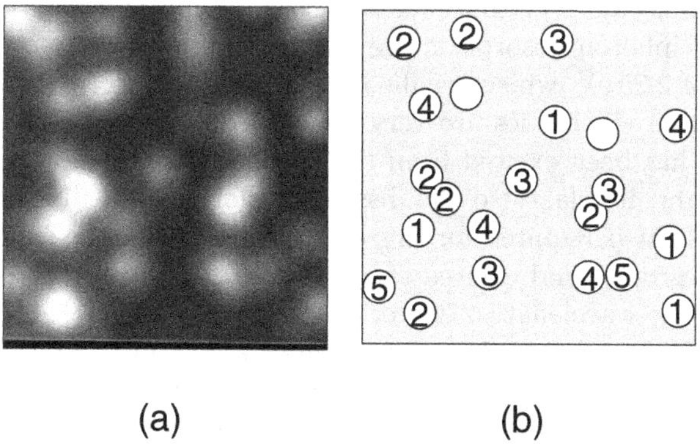

Fig. 3.12. Spatial distribution of the emission intensity. (**a**) Experimental results. Image size 3 μm × 3 μm. (**b**) Results of identifying electronic energies. Numbers 1–5 in *white circles* represent the orders of discrete electronic energies. Smaller numbers represent lower energies

electronic energies identified in the emission spectra, respectively. Such high resolution intensity mapping and spectral measurement for individual quantum dots has never been possible with conventional optical and spectroscopic methods.

It is not straightforward to achieve high spatial resolution using the experimental setup in Fig. 3.10 because the emission intensity is very weak. In order to overcome this difficulty, the high efficiency double-tapered-type fiber probe shown in Fig. 3.1d was developed. As a result, highly sensitive detection was achieved whilst maintaining high resolution [3.8].

As other examples, intensity mapping of emission from a single dye molecule (see Sect. 9.1.2) and Raman spectroscopy of organic/inorganic materials have been realized [3.15, 3.16].

3.3.3 Fabrication

This section reviews nanofabrication using the optical near field. This is a key application of the optical near field. Optical near-field photochemical vapor deposition (NFO-PCVD) was devised in order to fabricate nanometric materials by direct deposition on a substrate. In NFO-PCVD, the main process is the photochemical dissociation of a metal organic gas, which is performed by the optical near field. As an example, metallic zinc (Zn) deposition is illustrated schematically in Fig. 3.13a, where diethyl zinc molecules ($Zn(C_2H_5)_2$) are dissociated by the optical near field. A nanometric pattern of Zn can be deposited on the substrate surface using the photochemical reaction localized in the small space of the optical near field.

Figure 3.13b is a schematic diagram of the potential energy curves in a $Zn(C_2H_5)_2$ molecule. Thermal excitation or molecular vibronic excitations by infrared multiphoton absorption are required to go directly over the potential barrier of 2.25 eV, which is called the dissociation energy. However, since these excitation efficiencies are very low, pre-dissociation is preferred: after the electron has been excited from the ground state to the excited state by absorbing light, it relaxes to the dissociative orbital for dissociation. Ultraviolet (UV) light is required for this excitation because the energy difference between the ground and excited states is 4.59 eV, which corresponds to the absorption edge wavelength 270 nm.

The experimental deposition system uses a low pressure molecular gas of $Zn(C_2H_5)_2$ in a vacuum chamber, in which a UV fiber probe [3.7] is installed for illumination mode operation. The $Zn(C_2H_5)_2$ molecules are dissociated by the optical near field on the probe tip, and dissociated Zn atoms are deposited on a substrate. If the position of the fiber probe is fixed, the size and profile of the deposited atoms are determined by the spatial distribution of the optical-near-field energy. If the probe is scanned, various patterns can be deposited whose profiles depend on the scanning trajectory.

Figure 3.14a shows two hemispherical nanometric Zn dots deposited by fixing the fiber probe at two closely spaced positions [3.17]. The measured diameters of these dots are as small as 37 nm and 52 nm, where it should be noted that the real value is smaller than these values because the measured value is increased in a way that depends on the resolution of the measure-

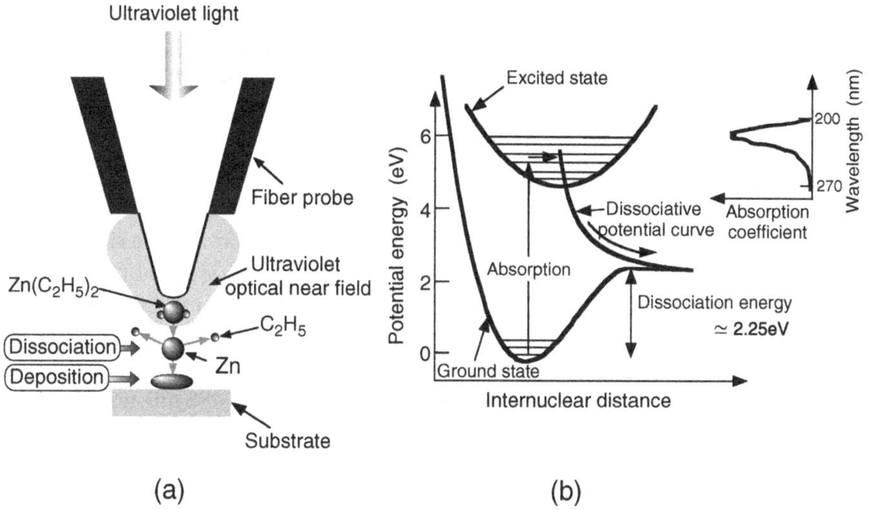

Fig. 3.13. Schematic illustration of photochemical vapor deposition of Zn by the optical near field. (**a**) Configuration of an experimental setup. (**b**) Schematic diagram of the potential energy curves in a $Zn(C_2H_5)_2$ molecule

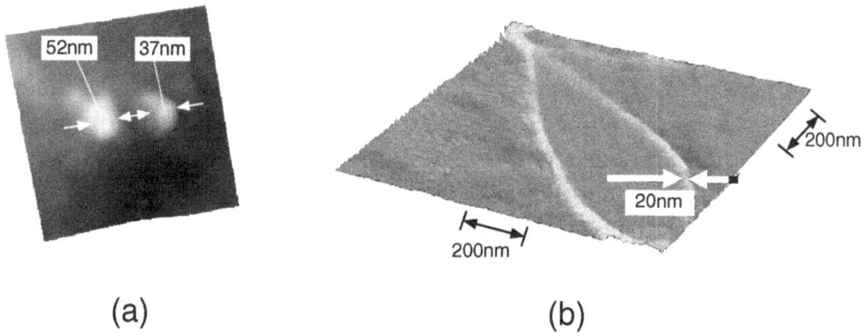

Fig. 3.14. Images of Zn deposited on a glass substrate. (**a**) Two closely deposited hemispherical Zn dots. (**b**) Elliptical pattern of a Zn wire

ment.[9] Figure 3.14b shows the elliptical pattern of the nanometric Zn wire deposited by scanning the fiber probe [3.18]. The measured width is as narrow as 20 nm, which is more than ten times narrower than the wavelength of the incident light (244 nm). It should also be noted that the real width is narrower than the measured value, for the reason given above.

Various sizes and profiles of deposited material can be obtained by controlling the fiber probe. Furthermore, the position of the deposited materials

[9] In order to obtain topographic images of these patterns, the magnitude of the shear force was mapped (see Sect. 3.3.1) with the fiber probe used for deposition. That is, a shear force microscope was used for imaging. This is a member of the category of atomic force microscopes reviewed in the supplement to Sect. 3.3.1.

can be accurately controlled. As shown in Fig. 3.14a, the separation between the two dots has been decreased until they are as close as their diameters. This is an outstanding advantage which has never been equalled by conventional deposition technologies. It thus shows great potential for fabricating nanometric materials with high reproducibility.

One advantage of using the optical near field for photochemical vapor deposition is the ability to fabricate nanometric patterns beyond the diffraction limit, as mentioned above. However, it should be noted that a more significant advantage is the availability of a unique mode of dissociation and deposition which has never been realized using conventional propagating light. For example, it has been confirmed experimentally that Zn is deposited even if 244 nm wavelength light is replaced by 488 nm wavelength light [3.19]. In general, it is not possible to dissociate $Zn(C_2H_5)_2$ using 488 nm wavelength light because the wavelength is much longer than the absorption edge of the molecule. This means that such low energy photons as are provided by 488 nm wavelength light cannot excite electrons in the molecule into the dissociative orbital.

For the optical near field, however, this unusual dissociation behavior is possible due to the intrinsic nature of the optical near field. In the case of conventional dissociation by propagating light, only the electron can respond to the oscillatory optical field while the nuclei do not, i.e., the process of dissociation is adiabatic. This adiabatic approximation is valid because the electric field of the propagating light is spatially homogeneous due to its sufficiently long wavelength compared with the molecular size. However, the optical near field is spatially inhomogeneous because of its very short decay length. Thus, the process becomes non-adiabatic, i.e., the nuclei in the molecule can respond to the optical near field. As a result, a unique dissociation process can occur.[10]

For example:

1. an electron can be directly excited from the ground state to the dissociative orbital due to non-adiabatic deformation of molecular orbital potentials,
2. high energy phonons in a $Zn(C_2H_5)_2$ molecule can be excited, although they cannot be coupled with propagating light.

The novel mode of dissociation described above has several technical advantages in addition to its profound scientific significance. For example, an expensive ultraviolet light source can be replaced by a conventional visible

[10] With regard to this non-adiabatic process, it should be noted that the main aim in applying the optical near field to novel areas of nanophotonics and atom photonics is to bring out and use these intrinsic phenomena and functions, which are not possible with propagating light. The nanophotonic switch in Sect. 9.3 is one example exploiting this intrinsic phenomenon. Compared with this primary objective, the ability to make nanometric measurements and fabrications are no more than secondary advantages.

light source. Further, since this dissociation is possible for various metal–organic molecular gases, it opens the way to depositions of novel materials that have never been realized using propagating light.

Besides Zn, metallic Al has been deposited. In addition, ZnO has been deposited. This is an oxide that can emit blue light. Not only metals and oxides, but composite semiconductors can also be deposited. This possibility of depositing a wide range of materials is a great advantage if we seek to fabricate nanophotonic integrated circuits with sizes much smaller than the light wavelength. This is because various nanometric materials can be deposited on the substrate with very high accuracy and reproducibility with regard to their size and position. Such novel nanophotonic integrated circuits can meet the requirements of the future society as described in Sect. 1.2.2.

Besides the above-mentioned deposition, several modes of fabrication have been demonstrated such as photolithography [3.20] and pulsed laser ablation [3.21], which may be able to meet the requirements of the future society described in Sect. 1.2.3.

3.3.4 Optical Disk Memory

Storage and readout of optical disk memory come under the heading of nanofabrication reviewed in the previous section. However, following the stage of basic study, the development of practical systems has begun. Indeed, extensive experimental work is under way to:

- write sub-wavelength-sized pits by modifying the surface of a magneto-optical medium or phase-change medium with the optical near field;
- read out written pits using illumination-mode near-field optical microscopy.

Since scanning with a fiber probe is not a fast way to achieve data transmission in the readout context, a two-dimensional array of silicon pyramidal probes has been employed. This so-called storage/readout head is explained schematically in Fig. 3.15. Figure 3.16 shows the profile of a contact slider fabricated by assembling this head and pads on a silicon substrate. The device slides over the storage medium which is coated with a thin lubricant film in order to maintain fast and stable sliding. A pit of length 110 nm has been stored and read out by sliding this device over a phase-change medium at sliding speeds as fast as 0.43 m/s [3.9]. As this result demonstrates, a prototype of practical near-field optical storage systems can be developed to realize high storage densities beyond the diffraction limit.

3.3.5 Extending Applications: Toward Atom Photonics

Besides the topics reviewed in Sects. 3.3.1–3.3.4, the optical near field has been applied to imaging, spectroscopy, and fabrication for various organic, inorganic, and biological samples. Commercial near-field optical microscopes

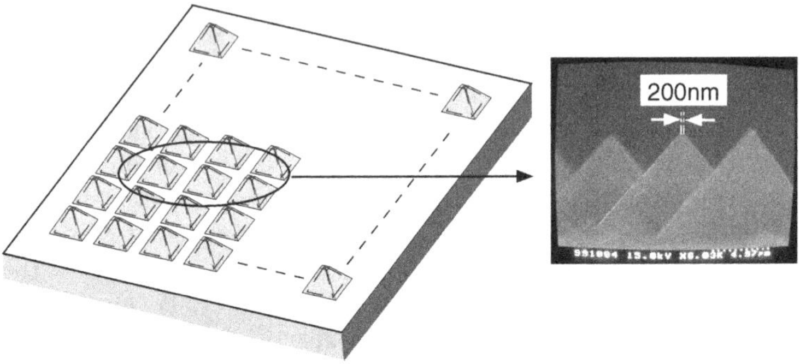

Fig. 3.15. Two-dimensional array of silicon pyramidal probes. *Left*: Schematic explanation of the structure. *Right*: SEM image

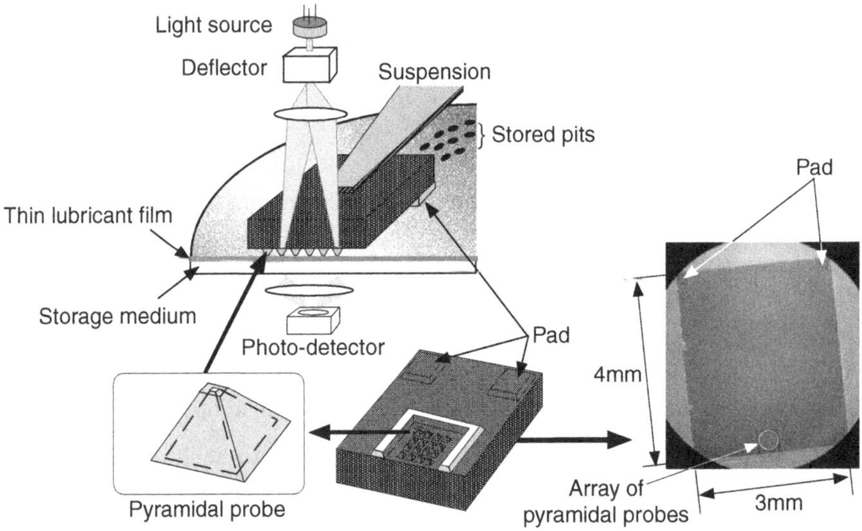

Fig. 3.16. Schematic representation of the experimental setup for high-density storage and high-speed readout using the optical near field

and near-field optical spectrometers are already available, and this should help to expand the field of applications in the future.

Atom manipulation has been proposed as an ultimate application of the optical near field, where the thermal motions of free neutral atoms in vacuum are controlled by using the mechanical effect of the optical near field. Several experiments have already succeeded in opening a new field of atom photonics, which is the next generation of nanophotonics. This section reviews recent progress in atom photonics, starting with the basic principles.[11]

[11] Refer to Sect. 9.2 for detailed explanations

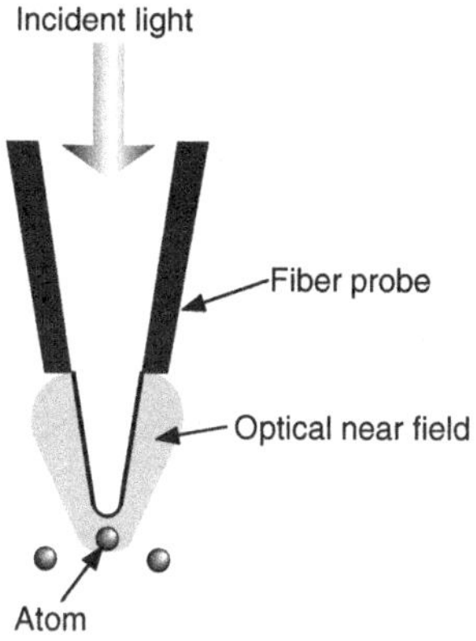

Fig. 3.17. Schematic diagram showing how to trap an atom by the optical near field on a probe tip

The idea is to use the dipole force, which is the force exerted by the electric field of light on the electric dipole of the atoms induced by the light. If the atom feels this force in the optical field and the optical frequency ν is lower than the resonant frequency ν_0 of the atom, it is forced to move toward the higher energy side of the optical field. On the other hand, the atom moves toward the lower optical energy side if $\nu > \nu_0$. Since the optical-near-field energy decreases with increasing separation from the material surface on which the optical near field is generated, the atom is attracted to the material surface by this dipole force if $\nu < \nu_0$. On the other hand, the dipole force becomes repulsive if $\nu > \nu_0$.

The possibility of confining an atom in the optical near field on a probe tip has been proposed, as explained schematically in Fig. 3.17 [3.22]. The figure demonstrates the possibility of atom trapping by the optical near field. The atom jumps into the optical near field on the fiber probe tip and is trapped by the balance between the attractive van der Waals force on the tip surface and the repulsive dipole force of the optical near field for $\nu > \nu_0$. Only one atom is trapped because of the small volume of the optical near field.

As a preliminary experiment, atom guidance has been carried out using a hollow optical fiber, as shown in Fig. 3.18a [3.23]. A cylindrical optical near field is generated on the inner surface of the hollow fiber by guiding a light through the core with doughnut-shaped cross-section. The atoms jump into the entrance of the hollow fiber, and are guided to the outlet by the repulsive force from the optical near field for $\nu > \nu_0$. However, the optical-near-field energy must be high enough to exert a stronger repulsive dipole force on the atom than the attractive force of adsorption onto the inner surface.[12]

Atom guidance experiments have been successfully carried out using rubidium (Rb) atoms. By heating metallic Rb to 200°C in an oven, Rb atomic

[12] This attractive force is the sum of the van der Waals force (see the supplement to Sect. 3.3.1) and the Casimir–Polder force, which has been called the cavity potential. (A detailed explanation is given in Sect. 5.3.

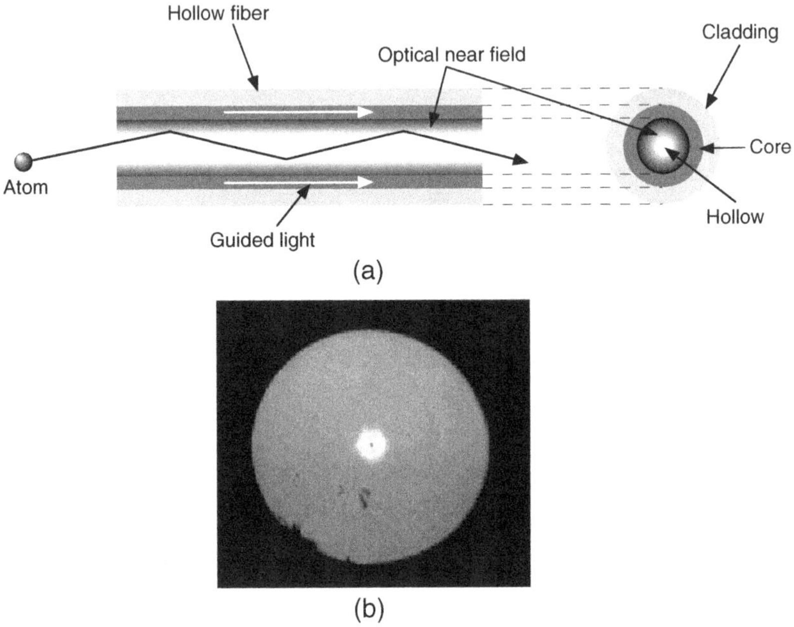

Fig. 3.18. Atom guidance by a hollow optical fiber. (**a**) Principle. (**b**) Cross-sectional profile of the hollow fiber. A *black circle* at the center represents the hollow with diameter 2 µm. The *white doughnut* around the black circle is the core. The larger *gray circle* is the cladding

vapor sprays into the high vacuum chamber with an average thermal atomic speed of about 300 m/s. The atoms are introduced into the hollow fiber. Figure 3.18b shows the cross section of a hollow fiber with inner diameter 2 µm. Several hollow fibers with diameters 0.3–0.7 µm and length 3 cm have also been used. The minimum diameter of 0.3 µm is smaller than the wavelength (780 nm) of the light entering the core.

Figure 3.19a represents the relation between the frequency difference $\nu - \nu_0$ and the measured flux of guided Rb atoms. The broken line parallel to the horizontal axis represents the value of the atom flux, which is transmitted through the hollow without an optical near field. This means that a small fraction of the atoms can transmit in a ballistic manner. As compared with this line, the solid curve represents a drastic increase in the guided flux for $\nu > \nu_0$, due to the repulsive dipole force of the optical near field. The value of the flux on this curve is lower than that on the broken line for $\nu < \nu_0$, due to the attractive dipole force in addition to the above-mentioned adsorptive force.

Figure 3.19b shows the relation between the light power introduced into the fiber core and the measured flux of guided Rb atoms, where the condition $\nu > \nu_0$ is maintained. The area around the origin of this relation is magnified and displayed in the inset. This inset shows that the atoms are not guided

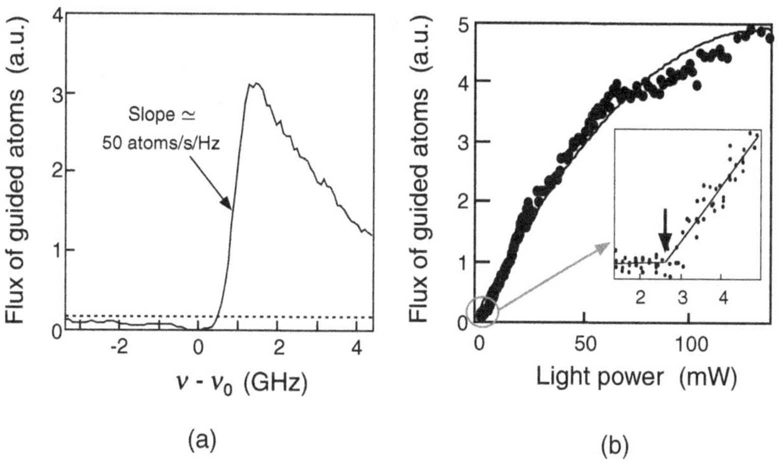

Fig. 3.19. Experimental results of Rb atom guidance. (**a**) Relation between the frequency difference $\nu - \nu_0$ and the measured flux of guided Rb atoms. The *broken line* parallel to the horizontal axis represents the value of the atom flux transmitted through the hollow without an optical near field. (**b**) Relation between the light power introduced into the fiber core and the measured flux of guided Rb atoms. The *black arrow* in the *inset* of this figure represents the guidance threshold

by the low light power, because the repulsive dipole force is weaker than the adsorptive force. The black arrow in the inset represents the threshold for guidance, i.e., the atom is guided when the light power is higher than this threshold value. Above the threshold, the flux of guided atoms increases with increasing light power and finally saturates, implying that atoms entering the hollow are totally guided.

This method can be applied not only to Rb but also to most of the atoms in the periodic table, if one can prepare a light source with emission frequency around the value of ν_0 for the guided atom. Table 3.1 shows typical atoms and their resonance wavelengths in vacuum, which are inversely proportional to ν_0. Several popular atoms can be seen in this table, e.g., Si and Ga, which are used in the electronics industry. It should be noted that the values of ν_0 in this table lie between the ultraviolet and near-infrared regions. Since lasers emitting coherent light in these regions have already been developed, this method can be applied to a variety of atoms from now onwards.

The atom guidance experiments may be considered as a preliminary study to the one in Fig. 3.17. However, a variety of applications have been found. For example, by noting that atom guidance is based on a resonance interaction between the optical near field and an atom, species of guided atoms can be selected, as shown in Table 3.1. Further, by tuning the optical frequency ν within a narrower range than that given in this table, isotope separation, control of atom flux, and atom deposition become possible. This will be reviewed below.

Table 3.1. Examples of atoms to be guided by the optical near field. Their resonance wavelengths are also given

Atom	Resonance wavelength [nm]	Atom	Resonance wavelength [nm]
Ag	328.1	K	766.5
Al	394.4	Li	670.8
Au	267.6	Mg	285.2
B	249.7, 249.8	Na	589.0
Be	234.9	O	777.5
Ca	422.7	Pb	368.4
Cs	852.1	Rb	780.0
Cu	327.4	Si	252.4
Ga	403.3	Sr	460.7
Hg	253.7		

Isotope Separation

Natural Rb contains two isotopes, i.e., ^{85}Rb and ^{87}Rb with the fractional ratio of 7:3. Their resonance frequencies ν_0 differ because of the difference in their masses, i.e., the value of ν_0 for ^{85}Rb is about 1 GHz higher than that of ^{87}Rb. By fixing the value of the laser frequency ν in-between these two values, the dipole force due to the optical near field is attractive for ^{85}Rb, while it is repulsive for ^{87}Rb. Thus, only the ^{87}Rb is guided through the hollow fiber while ^{85}Rb is adsorbed onto the hollow surface. This is the principle of isotope separation.

Curves A and B in Fig. 3.20 represent the spectral emission profiles of ^{85}Rb and ^{87}Rb measured at the exit and entrance of the hollow fiber, respectively. Comparison of these curves shows that the spectral peak of ^{85}Rb on curve A is much lower than that on curve B. This represents the fact that ^{85}Rb was adsorbed onto the hollow surface by the attractive dipole force.

This method of isotope separation is applicable to a variety of isotopes. One interesting example may be separation of carbon isotopes, which have been used as a medical tracer for diagnosing respiratory diseases. For this purpose, CO_2 molecules have to be guided through the hollow fiber. Intensive research into controlling the thermal motion of gaseous CO_2 molecules by the optical near field may open a new field of carbon isotope separation in the future.

Atom Flux Control and Atom Deposition

Isotope separation reviewed above was realized by tuning the laser frequency ν with respect to the atomic resonance frequency ν_0. Since the value of ν was accurately controlled to reduce its fluctuations to less than 100 kHz, it

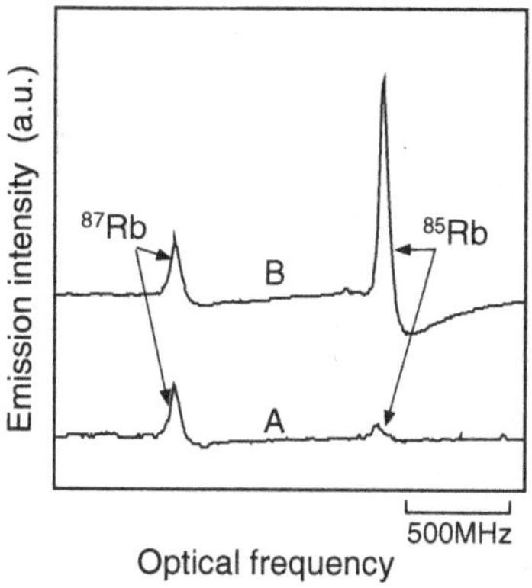

Fig. 3.20. Experimental results of isotope separation of Rb atoms. (**a**) Curves A and B represent the spectral emission profiles of ^{85}Rb and ^{87}Rb measured at the exit and entrance of the hollow fiber, respectively

is possible to fix ν at one point on the linear part of the curve in Fig. 3.19a, which has a slope of 50 atoms/s Hz. From the product of the magnitude of frequency fluctuations and the slope, the inaccuracy in the guided atom flux can be made as low as 5×10^{-3} atoms/s, which means that very accurate atomic flux control is possible.

Such high accuracy also suggests the possibility of atom deposition by spraying guided atoms onto a cold substrate. Preliminary experiments have estimated that the deposition rate is 0.2 nm/min. Further experiments are in progress and a new field of atom photonics is expected to develop.

3.4 New Areas of Optical Science Exploiting Optical Near Fields

The topics reviewed in the previous sections suggest the possibility of developing a new technology of nanophotonics and even atom photonics using the optical near field. However, apart from these new technologies, a new field of science can also be created by studying the nature of the optical near field. For this study, it is necessary to investigate the elementary process of local electromagnetic interaction between light and matter in a space of sub-wavelength dimension. For example, as was shown in Fig. 2.6b, the approach of the sphere P varies the orientations and positions of the electric dipole moments in the sphere S, so that the distribution of electric lines of force varies from its original configuration, as shown in Fig. 2.3. In contrast to these variations, note that the refractive index is an intrinsic physical constant in conventional optics, determined by the orientations and positions of the electric dipole moments in matter. However, since the above-mentioned variations mean that the refractive index of the sphere S is no longer an invariant physical constant, a unique approach is required to describe the optical properties of this matter, differing from the theory of conventional optics.

On the other hand, as described at the end of Sect. 2.1, the description of optical-near-field generation involves a hierarchical model, which is a unique feature of the mesoscopic scale. Since the interaction between light and matter in the mesoscopic range takes place within a surrounding macroscopic system, a novel theoretical model is required to describe the microscopic system in a macroscopic heat bath.[13] Furthermore, novel interface and device concepts are required to transfer the optical near field energy generated in the microscopic system to the macroscopic system.[14]

In relation to the comments given above, several basic problems have been pointed out, e.g., the problem of regarding optical-near-field detection as photon tunneling, the problem of signal transmission utilizing optical energy transfer from the light source to the photodetector via the sample and probe, and so on. Since they will be discussed in Sect. 9.3, the remainder of this section presents some comments concerning application of the optical near field to measurement and fabrication [3.24]. This will serve as an introduction to the theories presented in Chaps. 4–8.

First, it should be noted that the system composed of the spheres S and P and the optical near field has nanometric size, much smaller than the wavelength of the light and much larger than the size of atoms. Further, the light and matter are mutually dependent because the spheres are connected by electric lines of force of the optical near field. Therefore, one has to study the combined system of light and matter in the mesoscopic region, as described at the end of Sect. 2.1. Such light–matter interaction and its applications has never been studied in conventional optical science. An especially important point is that the information required is obtained by carefully filtering the detected signals. Actually, in the case of near-field optical microscopy, the shape and size of the fiber probe are designed to fulfill the conditions of an optimum band-pass filter (see Fig. 3.2b), so as to obtain a sample image with high resolution and contrast. Such filtering is a method inherent to mesoscopic science and technology. In contrast, conventional optics observes the sample only in the far field, after passing through a low-pass filter due to the diffraction of light.

Second, the nature of the optical near field is discussed with regard to the optical response of matter. For example, Fig. 2.3 shows that light illumination generates electric dipole moments in a dielectric, and that these are connected by the generated electric lines of force. However, this connection is broken at the surface of the matter, whereby polarized electric charges are induced to generate the optical near field. Since the dielectric must be electri-

[13] This theoretical model is presented in Chap. 8.

[14] As described in Sect. 3.2, the tapered part of the fiber probe plays the role of interface device. In addition, the second requirement mentioned there corresponds to the function of interfacing.

cally neutral, the rear surface induces polarized electric charges with opposite sign. As a result, these charges generate an electric field in a direction opposite to the electric field of the incident light. Therefore, the magnitude of the local electric field inducing electric dipole moments in the matter is the difference between the electric field of the incident light and that generated from the surrounding electric dipole moments. Thus, the magnitude of the electric dipole moment in the matter is determined by the internal electromagnetic interaction, i.e., the sum of the local response of the electric dipole moment to the incident light and contributions from surrounding electric dipole moments. This means that the magnitude of the response of the matter to the incident light is obtained if the spatial distribution of induced electric dipole moments and the electric field are determined in a consistent manner. Such a consistently determined electric field is called a self-consistent field. Even the nonlocal optical response can be observed if one observes part of the matter whose size is as small as the spread of the electron wave function [3.25].

Third, the hierarchy presented at the end of Sect. 2.1 is discussed again, i.e., the hierarchy into which the optical response is organised by interactions in the matter. For this discussion, Fig. 3.21a illustrates the electric dipole moments generated by illuminating a nanometric material with incident light. Dividing this matter into parts A and B as shown in Fig. 3.21b, the optical response of this matter to the incident light can be expressed as the sum of the contributions from parts A and B, and the contribution from the interaction between A and B. A and B generate self-consistent fields depending on their shapes and sizes. Here, the sum of their magnitudes is smaller than that of the self-consistent field of the undivided matter (i.e., A+B) because of the interaction between A and B. It is thus found that the optical response of matter has a hierarchy, i.e., the optical response of a larger piece of matter is the sum of the responses of the smaller pieces and their interactions. Therefore, various characteristics of the interaction can be observed, depending on how one divides the matter up. An iterative method of calculation has been proposed to derive the magnitude of the self-consistent field by summing the magnitudes of the optical responses and interactions between homogeneously divided smaller parts [3.26].

Detection of the optical near field corresponds to extracting specific interactions by bringing the probe towards the sample instead of dividing the sample, as shown in Fig. 3.21b. That is, the spheres S and P correspond to the parts A and B in Fig. 3.21b. This means that detection of the optical near field amounts to inducing an interaction between the spheres S and P. Measurement and fabrication beyond the diffraction limit can be realized by exploiting this interaction. In other words, interaction between spheres S and P can be extracted by measuring $I_{S+P} - I_S - I_P$, where I_{S+P}, I_S, and I_P are the light intensities scattered from S + P, S, and P, respectively. If the shape and size of the probe P are optimized, and if control of its position is also optimized, the interaction between S and P is correlated to the optical near field

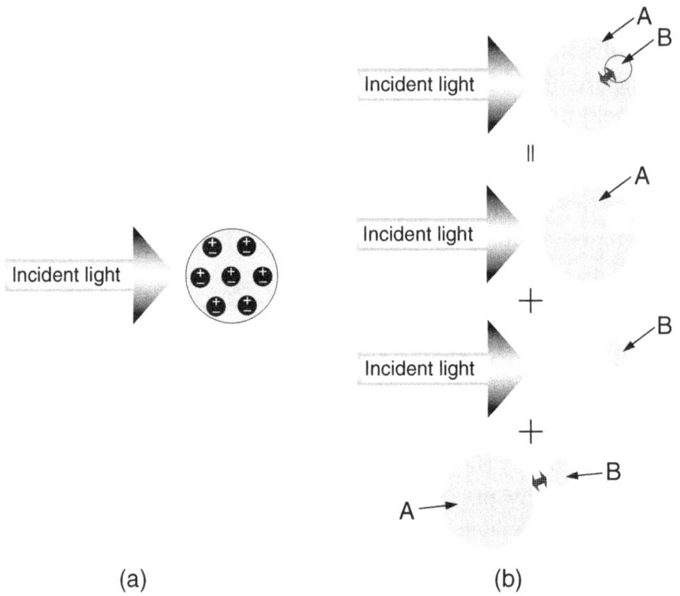

Fig. 3.21. Schematic explanation of hierarchy. (**a**) Electric dipole moments in a nanometric material, generated by the incident light. (**b**) Optical response to the incident light as a sum of contributions from parts A and B, and a contribution from the interaction between A and B

on the surface of S which existed before the approach of P. Thus, the optical near field is reconstructed by analyzing the optical responses measured while scanning P. This is the operation principle of a near-field optical microscope.

As described above, the optical near field is detected through the near-field electromagnetic interaction induced by bringing P towards S. In order to use the optical near field for measurement and fabrication, it is necessary to induce this interaction efficiently, to collect the scattered light selectively, and to increase the contrast between the collected signal and the background signal. This requirement can be met by optimizing the shape, size, and position of the probe P, depending on the properties of S. This optimization corresponds to adjusting the characteristics of the band-pass filtering, as shown in Fig. 3.2b. In particular, all the light except the optical near field under consideration must be screened to achieve high contrast. This can be done using the screen in Fig. 2.7 or the fiber probe in Fig. 2.8.

It is also possible to enhance the optical-near-field energy by a resonance effect of matter. Since electronic systems in a nanometric material exhibit various quantum effects, a novel probe can be realized exhibiting the above-mentioned band-pass filtering and resonance characteristics [3.8, 3.11]. Further, novel transitions and interactions can be induced by the optical near field such as are inhibited in macroscopic systems [3.25, 3.27]. One such novel transition has already contributed to photochemical vapor deposition (see Sect. 3.3). Thus, the interaction between light and electronic systems in the mesoscopic region has great scope for versatile applications.

Further, it should be noted that light can have a mechanical action on matter originating in its electric and magnetic components. The former is due to the transfer of photon momentum to matter. This transfer has been applied to control the thermal motion of atoms, as described in Sect. 3.3.5. The latter is due to the transfer of photon angular momentum to the orbital angular momentum of electronic systems and due to the spin–orbit interaction of electronic systems.

Since the optical near field is coupled to matter, definitions of momentum and angular momentum as well as their conservation laws become approximate and depend strongly on the shape and size of the matter. Therefore, these momenta are called pseudo-momenta. For example, in the case of conventional propagating light in free space, the direction of its intensity gradient is parallel to that of the wave number, while they are perpendicular for the optical near field.[15] Further, since it is possible for the localized optical near field to exert a large mechanical force on the matter, a novel mechanical phenomenon inherent in the optical near field can be expected. Experimental work to measure such mechanical effects has been carried out [3.28].

It is known that a local interaction can be induced between closely spaced pieces of matter which originates from the polarization of matter and vacuum fluctuations. It is induced even in the absence of the incident light, and hence generates a van der Waals force (see Sect. 5.3). In contrast to this, the mechanical action of the optical near field can be considered as an atomic force which can be controlled by adjusting external parameters. Applications of such mechanical action intrinsic to the optical near field include fabrication, control of thermal motion of atoms, and others, as described in the present chapter. A variety of other applications, e.g., manipulation of biological specimens, also become possible.

Problems

Problem 3.1

Is it advantageous to use a diffraction grating as a standard sample to evaluate the resolution of a near-field optical microscope?

Problem 3.2

A fiber probe is used for the illumination mode in order to generate the optical near field on the probe tip. Is high resolution obtained if the light scattered from the sample is collected by a convex lens when the sample is illuminated by the optical near field?

[15] For example, refer to (2.6) in the supplement to Sect. 2.1.

4 Dipole–Dipole Interaction Model of Optical Near Field

Although one can acquire some information about the optical near field by solving the approximated Maxwell and Schrödinger equations simultaneously, this derivation requires numerical calculations with a very long computation time. Even though numerical results are possible, it is very difficult to obtain an intuitive physical picture of the optical near field. In order to overcome this difficulty, Chaps. 4–9 seek to review theoretical models that offer intuitive concepts to analyze the physical meaning of the optical near field and the relevant experimental results. The present chapter presents the simplest theoretical model to describe the phenomena presented in Chap. 2. Section 4.1 imposes a condition on the size of material systems in which the optical near field is investigated. Under this condition, Sect. 4.2 describes the basic role of a probe from the viewpoint of the dipole–dipole interaction. Section 4.3 discusses the characteristics of fiber probes, which depend on their shape and composition.

4.1 Near-Field Condition for Detecting Scattered Light

This section treats light scattering by a small object. For simplicity, the material object is assumed to be composed of two particles, as shown in Fig. 4.1. Their separation $\boldsymbol{b}$ can be regarded as the size of the material object. In this figure, the vector $\boldsymbol{r}$ represents the distance between the detection point and material object. The following equations are used to determine the value of b $(= |\boldsymbol{b}|)$ from the scattered light intensity measured at $\boldsymbol{r}$. Since the two particles can be regarded as point light sources for light scattering, the electric field vector $\boldsymbol{E}(\boldsymbol{r}, t)$ of the scattered light at time t and the position $\boldsymbol{r}$ is expressed as

$$\boldsymbol{E}(\boldsymbol{r}, t) = \boldsymbol{E}_0 \frac{\mathrm{e}^{-\mathrm{i}\omega t + \mathrm{i}k|\boldsymbol{r}+\boldsymbol{b}/2|}}{|\boldsymbol{r}+\boldsymbol{b}/2|^m} + \boldsymbol{E}_0 \frac{\mathrm{e}^{-\mathrm{i}\omega t + \mathrm{i}k|\boldsymbol{r}-\boldsymbol{b}/2|}}{|\boldsymbol{r}-\boldsymbol{b}/2|^m} , \tag{4.1}$$

where $\boldsymbol{E}_0$, ω and k are the electric field vector of the incident light, the angular frequency $(= 2\pi\nu)$ and the wave number $(= 2\pi/\lambda$, where λ is the wavelength) of the incident light, respectively. Since the travel time Δt of the light from the light source to the detection point is given by $|\boldsymbol{r} \pm \boldsymbol{b}/2|/c$,

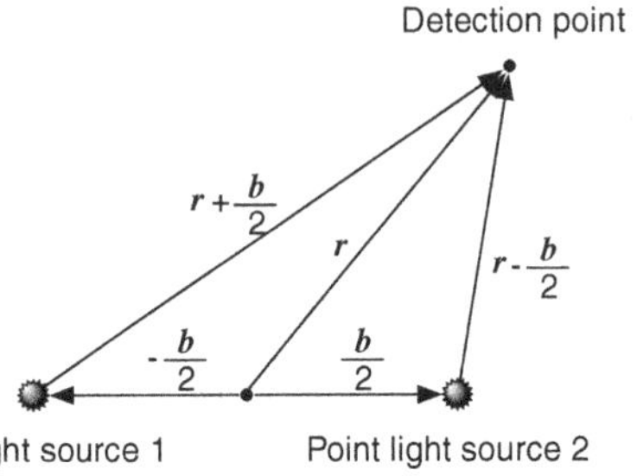

Fig. 4.1. Configuration for optical measurement of the separation b between two point light sources

the quantity $k|\boldsymbol{r} \pm \boldsymbol{b}/2|$ in this equation represents the phase delay $\omega\Delta t$. The integer m represents the exponent originating from (A.28) of Appendix A, which takes values 1, 2, or 3. Three possible cases can be considered to derive the value of b, depending on the values of b and $r(= |\boldsymbol{r}|)$.

Case 1: $1 \ll kb \ll kr$

In this case we are observing a super-wavelength-sized object in the far field. Since $1 \ll kr$, the term proportional to r^{-1} in (A.28) is larger than those proportional to r^{-2} and r^{-3}. The ratio of their magnitudes is $(kr)^2 : kr : 1$. Hence, the value of m in (4.1) can be fixed at unity. Further, by noting that $r \gg b$, (4.1) approximates to

$$\boldsymbol{E}(\boldsymbol{r}, t) \approx 2\boldsymbol{E}_0 \frac{\mathrm{e}^{-\mathrm{i}\omega t + \mathrm{i}kr}}{r} \cos \frac{kb(\boldsymbol{n} \cdot \boldsymbol{n}_b)}{2} , \tag{4.2}$$

where $\boldsymbol{n}(= \boldsymbol{r}/|\boldsymbol{r}|)$ and $\boldsymbol{n}_b(= \boldsymbol{b}/|\boldsymbol{b}|)$ represent unit vectors oriented along $\boldsymbol{r}$ and $\boldsymbol{b}$, respectively. The symbol $\cdot$ represents the scalar product of the two vectors. In order to estimate the value b from this equation, assume that $|\boldsymbol{E}(\boldsymbol{r}, t)|$ has maximum value at the position $\boldsymbol{r}_0$ at which $(\boldsymbol{n}_0 \cdot \boldsymbol{n}_b) = 0$ ($\boldsymbol{n}_0 = \boldsymbol{r}_0/|\boldsymbol{r}_0|$, see Fig. 4.2). If one can find the second position $\boldsymbol{r}_1$ at which $|\boldsymbol{E}(\boldsymbol{r}, t)|$ has the maximum value again while the value $|\boldsymbol{r}|$ is constant, the relation $kb(\boldsymbol{n}_1 \cdot \boldsymbol{n}_b)/2 = \pi$ is valid, and $\boldsymbol{n}_1 = \boldsymbol{r}_1/|\boldsymbol{r}_1|$. From this relation, the value of b is derived as $b = 2\pi/k(\boldsymbol{n}_1 \cdot \boldsymbol{n}_b)$.

The reason why the value of b can be obtained as above is that the product kb is much larger than unity, i.e., the detection point is located sufficiently far from the two light sources, so that the phase difference of the light waves traveling from the two light sources is larger than π. Imaging by a conventional optical microscope is a case in point.

Case 2: $kb \ll 1 \ll kr$

In this case we are observing a sub-wavelength-sized object in the far field. The value of m in (4.1) takes the value unity again in this case. In or-

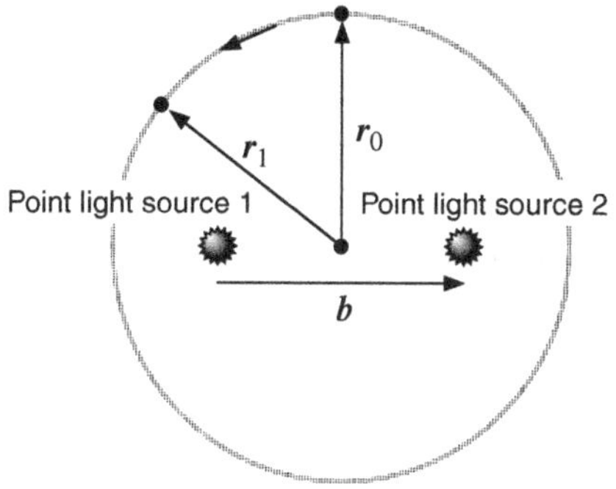

Fig. 4.2. $|\boldsymbol{E}(\boldsymbol{r},t)|$ reaches its maximum at points $\boldsymbol{r}_0$ and $\boldsymbol{r}_1$

der to estimate the value b from (4.2) under this condition, the inequality $kb|\boldsymbol{n}\cdot\boldsymbol{n_b}| \geq 2\pi$ has to be satisfied, i.e., at the detection point the phase difference between the light waves of the two light sources must be larger than π. Since $|\boldsymbol{n}\cdot\boldsymbol{n_b}| \leq 1$, the above requirement corresponds to $kb \geq 2\pi$, which represents the diffraction limit. However, this requirement is not met because $kb \ll 1$, and thus the value of b cannot be obtained, i.e., the spatial distribution of the position of the sub-wavelength-sized object cannot be measured in the far field. This is because the phase difference between the two light waves is negligibly small at any detection point $\boldsymbol{r}$. Rayleigh scattering is a case in point.[1]

Case 3: $kb < kr \ll 1$

In this case we are observing a sub-wavelength-sized object in the near field. Since $kr \ll 1$, the term proportional to r^{-3} in (A.28) is larger than those proportional to r^{-2} and r^{-1}. Their ratio is $(kr)^{-2} : (kr)^{-1} : 1$, and therefore $m = 3$. Further, since the phase delay $k|\boldsymbol{r} \pm \boldsymbol{b}/2|$ is negligibly small, (4.1) approximates to

$$\boldsymbol{E}(\boldsymbol{r},t) = \boldsymbol{E}_0 \mathrm{e}^{-\mathrm{i}\omega t}\left(\frac{1}{|\boldsymbol{r}+\boldsymbol{b}/2|^3} + \frac{1}{|\boldsymbol{r}-\boldsymbol{b}/2|^3}\right) . \tag{4.3}$$

We assume that the amplitude of the electric field measured at the detection point $\boldsymbol{r}_0$ is $|\boldsymbol{E}_1|$, and that $\boldsymbol{r}_0$ is normal to the line connecting the two particles, i.e., $(\boldsymbol{n}_0\cdot\boldsymbol{n_b}) = 0$. Then the value b is derived as

[1] It is a light-scattering phenomenon observed when the light is incident upon sub-wavelength particles. Equation (A.35) of Appendix A shows that the scattered light intensity is proportional to λ^{-4} or ν^4, where λ and ν are the wavelength and frequency of the incident light, respectively. Light scattering by gaseous molecules is an example of this phenomenon, and is the origin of the blue color of the daytime sky.

$$b = 2\left[\left(\frac{2|\boldsymbol{E}_0|}{|\boldsymbol{E}_1|}\right)^{2/3} - r_0^2\right]^{1/2},$$

from the relation $|\boldsymbol{E}_1| = 2|\boldsymbol{E}_0|/\left[r_0^2 + (b/2)^2\right]^{3/2}$. This means that the value of b can be determined by the near-field measurement. The relation $kb < kr \ll 1$ is called the near-field condition [4.1].[2] The topics to be discussed in this book are the phenomena observed under this condition.

4.2 Role of Probes

The optical near field is used for imaging, fabrication, manipulation, and so on, under the near-field condition. In this section a simple case is considered to elucidate the role of probes, namely, the case where both the sample and the probe are treated as spherical particles.

4.2.1 Strength of Dipole Interaction

Figure 4.3 shows the light incident on the sphere S used as a sample. Simultaneously, the sphere P used as a probe is also illuminated by the incident light because it is close enough to the sphere S. Therefore, we can discuss not only the collection mode, but also the illumination mode, using this figure.

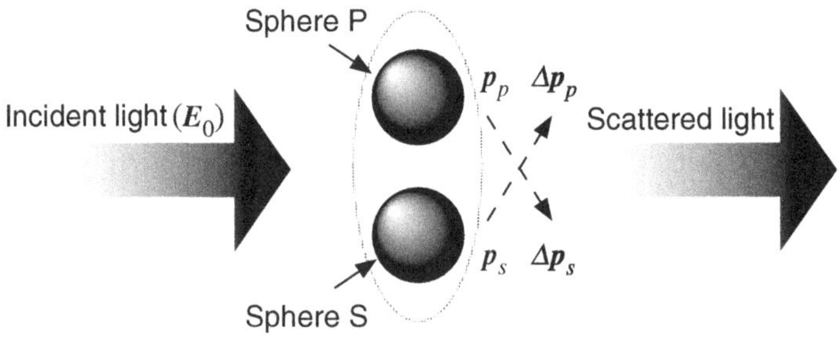

Fig. 4.3. Electric dipole moments induced in spheres S and P which are located close to each other

The electric field $\boldsymbol{E}_0$ of the incident light induces electric dipoles in the spheres, whose magnitude and orientation are represented by the electric dipole moments of (A.25) in Appendix A, and they are represented as $\boldsymbol{p}_S$

[2] The range of r satisfying the condition $kr \ll 1$ is called the near field. On the other hand, the one satisfying $kr \gg 1$ is called the far field.

and $\boldsymbol{p}_{\mathrm{P}}$ in spheres S and P, respectively.[3] The electric dipole moment $\boldsymbol{p}_{\mathrm{S}}$ in the sphere S generates an electric field which induces the change $\Delta\boldsymbol{p}_{\mathrm{P}}$ in the electric dipole moment of the sphere P. Similarly, the electric field generated by $\boldsymbol{p}_{\mathrm{P}}$ induces the change $\Delta\boldsymbol{p}_{\mathrm{S}}$ in the electric dipole moment of the sphere S. Further, $\Delta\boldsymbol{p}_{\mathrm{P}}$ and $\Delta\boldsymbol{p}_{\mathrm{S}}$ generate electric fields which induce changes in the electric dipole moments in the spheres S and P, respectively. The process will be repeated infinitely. This electromagnetic interaction, which mutually induces electric dipole moments in the two spheres, is called the dipole–dipole interaction [4.2]. All the electric dipole moments induced by this interaction can generate scattered light, which is the scattered light 2 of Fig. 2.6. We now elucidate the details of the dipole–dipole interaction.

Equation (A.28) in Appendix A represents the electric field $\boldsymbol{E}$ generated by the electric dipole moment $\boldsymbol{p}$. In the case of the near-field condition ($kr \ll 1$), the main contribution comes from the term proportional to r^{-3} in this equation. It is given by

$$\boldsymbol{E} = \frac{3\boldsymbol{n}(\boldsymbol{n}\cdot\boldsymbol{p}) - \boldsymbol{p}}{4\pi\varepsilon_0 r^3} , \tag{4.4}$$

where $\mathrm{e}^{\mathrm{i}kr}$ is approximated as unity because $kr \ll 1$. It corresponds to the static component of the electric field generated by the electric dipole moment.[4] When $\boldsymbol{r}$ is parallel to $\boldsymbol{p}$ (i.e, $\boldsymbol{r} \parallel \boldsymbol{p}$), (4.4) is expressed as

$$\boldsymbol{E} = \frac{2\boldsymbol{p}}{4\pi\varepsilon_0 r^3} . \tag{4.5}$$

On the other hand, when $\boldsymbol{r}$ is perpendicular to $\boldsymbol{p}$ (i.e., $\boldsymbol{r} \perp \boldsymbol{p}$), it is

$$\boldsymbol{E} = -\frac{\boldsymbol{p}}{4\pi\varepsilon_0 r^3} . \tag{4.6}$$

These equations represent the optical near field generated around the sphere S. The dipole–dipole interaction induced by bringing the sphere P towards the sphere S is used to measure its magnitude.

When we assume that the spheres S and P are dielectrics for simplicity, the electric dipole moment $\boldsymbol{p}_{\mathrm{S}}$ of S induced by the incident electric field $\boldsymbol{E}_0$ is expressed as

$$\boldsymbol{p}_{\mathrm{S}} = \alpha_{\mathrm{S}}\boldsymbol{E}_0 , \tag{4.7}$$

where α_{S} is the polarizability of the dielectric. When the separation R between the two spheres satisfies the conditions $kR \ll 1$ and $\boldsymbol{R} \parallel \boldsymbol{p}$, the

---

[3] The spheres shown in Figs. 2.3 and 2.6b correspond to the ensemble of these spheres. They contain large numbers of electric dipole moments. Hence, the discussion in the present section can be used to make the discussions in Sects. 2.1 and 2.2 more quantitative.

[4] In the case when the near-field condition does not hold, the terms proportional to r^{-2} and r^{-1}, and also the term $\mathrm{e}^{\mathrm{i}kr}$ in (A.28), have to be taken into account. More details are given in Sect. A.1.2.

electric field at the sphere P generated by $\boldsymbol{p}_S$ is expressed by (4.5) as $\boldsymbol{E}_S = 2\boldsymbol{p}_S/4\pi\varepsilon_0 R^3$.[5] Thus, the change in the electric dipole moment of the sphere P is given by

$$\Delta \boldsymbol{p}_P = \alpha_P \boldsymbol{E}_S = \frac{2\alpha_P \alpha_S}{4\pi\varepsilon_0 R^3}\boldsymbol{E}_0 \,. \tag{4.8}$$

By representing the change in (4.8) as $\Delta\alpha_P \boldsymbol{E}_0$, the change in the polarizability of the sphere P is given by[6]

$$\Delta\alpha_P = \frac{\alpha_P \alpha_S}{2\pi\varepsilon_0 R^3} \,. \tag{4.9a}$$

Here, α_S and α_P are given by

$$\alpha_i = g_i a_i^3 \,, \tag{4.9b}$$

$$g_i = 4\pi\varepsilon_0 \frac{\varepsilon_i - \varepsilon_0}{\varepsilon_i + 2\varepsilon_0} \,, \tag{4.9c}$$

for $i = \mathrm{S}, \mathrm{P}$, where a_S, a_P, ε_S and ϵ_P are the radii and dielectric constants of the spheres S and P, respectively.[7]

Even if the suffix S is replaced by P, the above discussion is still valid, i.e., the electric dipole moment $\boldsymbol{p}_P$ $(= \alpha_P \boldsymbol{E}_0)$ generates the electric field $\boldsymbol{E}_P$ $(= 2\boldsymbol{p}_P/4\pi\varepsilon_0 \boldsymbol{R}^3)$ at the position of the sphere S, and induces the change $\Delta\boldsymbol{p}_S$ $(\equiv \Delta\alpha_S \boldsymbol{E}_0)$ in the electric dipole moment. As a result, one finds that the expression for the change $\Delta\alpha_S$ in the polarizability is the same as (4.9a). Thus, $\Delta\alpha_S$ and $\Delta\alpha_P$ are given by

$$\Delta\alpha_S = \Delta\alpha_P = \Delta\alpha \,. \tag{4.10}$$

In the following discussion, multiple scattering is neglected for simplicity, i.e., further changes in the electric dipole moments of the spheres S and P, induced by $\Delta\boldsymbol{p}_P$ and $\Delta\boldsymbol{p}_S$, respectively, are neglected.

Since the two spheres are close enough to each other, they are recognized as a single scatterer for far-field detection, i.e., the scattered light generated

---

[5] The discussion below can be made similarly for the case $\boldsymbol{R} \perp \boldsymbol{p}$.

[6] In conventional optics, the polarizability α and refractive index n of a material object are regarded as invariant constants. They are related to each other by $n = \sqrt{1 + N\alpha/\varepsilon_0}$, where N and ε_0 are the numbers of electric dipole moments in a unit volume and the dielectric constant in vacuum, respectively. However, (4.9a) shows that the polarizability of the sphere P depends on the separation R as the result of dipole–dipole interaction with the sphere S. This means that the refractive index is no longer an invariant constant in near-field optics. (The boundary condition is also regarded as invariant in conventional optics, and this turns out to be effective in near-field optics as well. For a detailed discussion on the boundary condition, refer to Sect. 7.1.1.)

[7] For derivation, refer to Problem 4.3

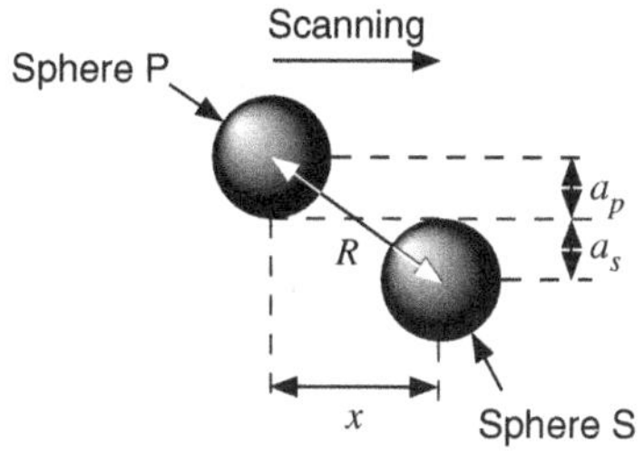

Fig. 4.4. Schematic diagram of spheres P and S with separation R

from the total electric dipole moment $\boldsymbol{p}_\mathrm{P} + \Delta\boldsymbol{p}_\mathrm{P} + \boldsymbol{p}_\mathrm{S} + \Delta\boldsymbol{p}_\mathrm{S}$ is detected. Its intensity I_S is given by

$$\begin{aligned} I_\mathrm{S} &\propto |(\boldsymbol{p}_\mathrm{P} + \Delta\boldsymbol{p}_\mathrm{P}) + (\boldsymbol{p}_\mathrm{S} + \Delta\boldsymbol{p}_\mathrm{S})|^2 \\ &\approx (\alpha_\mathrm{P} + \alpha_\mathrm{S})^2 |\boldsymbol{E}_0|^2 + 4\Delta\alpha(\alpha_\mathrm{P} + \alpha_\mathrm{S})|\boldsymbol{E}_0|^2 \,. \end{aligned} \tag{4.11}$$

The first term $(\alpha_\mathrm{P} + \alpha_\mathrm{S})^2 |\boldsymbol{E}_0|^2$ represents the intensity of the light scattered directly by the spheres S and P, which corresponds to the scattered light 1 of Fig. 2.1. The contribution of this term will be discussed in Sect. 4.2.3. The second term $4\Delta\alpha(\alpha_\mathrm{P}+\alpha_\mathrm{S})|\boldsymbol{E}_0|^2$ represents the intensity of the scattered light as the result of dipole–dipole interactions, which corresponds to the scattered light 2 of Fig. 2.6.

Combining (4.9) and (4.10) we obtain

$$\Delta\alpha = \frac{g_\mathrm{P} g_\mathrm{S}}{2\pi\varepsilon_0} \frac{a_\mathrm{P}^3 a_\mathrm{S}^3}{R^3} \,. \tag{4.12}$$

For convenience, the right-hand side of this equation is transformed as $(g_\mathrm{P} g_\mathrm{S}/2\pi\varepsilon_0) f(x)$, where the function $f(x)$ denotes $a_\mathrm{P}^3 a_\mathrm{S}^3/R^3$ and x is the horizontal component of the separation R. For simplicity, we assume that sphere P remains close to sphere S during scanning (see Fig. 4.4). Under this assumption, the orientations of the electric dipole moments in the two spheres can be maintained parallel to one another while scanning the sphere P, and thus (4.5) is valid.

Since the separation R is given by $\sqrt{x^2 + (a_\mathrm{P} + a_\mathrm{S})^2}$, $f(x)$ can be written by

$$f(x) \equiv \frac{a_\mathrm{P}^3 a_\mathrm{S}^3}{\left[x^2 + (a_\mathrm{P} + a_\mathrm{S})^2\right]^{3/2}} \,. \tag{4.13}$$

Figure 4.5 shows the profile of the function $f(x)$. The value of $f(x)$ reaches a maximum f_m at $x = 0$ (see Fig. 4.5), given by

$$f_\mathrm{m} = \frac{a_\mathrm{P}^3 a_\mathrm{S}^3}{(a_\mathrm{P} + a_\mathrm{S})^3} \equiv a_\mathrm{S}^3 \left(\frac{A_\mathrm{P}}{A_\mathrm{P} + 1}\right)^3 , \tag{4.14}$$

where $A_\mathrm{P} = a_\mathrm{P}/a_\mathrm{S}$. The following sections discuss the characteristics of several quantities derived from the above results.

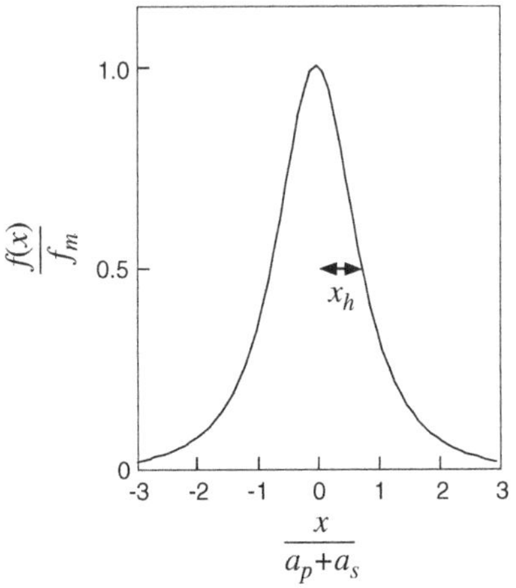

Fig. 4.5. Dependence of $f(x)$ on $x/(a_{\mathrm{P}}+a_{\mathrm{S}})$, where x_{h} is the half width at half maximum

4.2.2 Signal Intensity and Resolution

The signal intensity f_{m} of (4.14) increases with increasing a_{P} (see Fig. 4.6) and approaches a_{S}^3 for $a_{\mathrm{P}} \to \infty$. On the other hand, the half width x_{h} at the half maximum of the function $f(x)$ (see Fig. 4.5) is given by

$$x_{\mathrm{h}} = \sqrt{4^{1/3}-1}(a_{\mathrm{P}}+a_{\mathrm{S}}) \approx 0.77(a_{\mathrm{P}}+a_{\mathrm{S}}) = 0.77a_{\mathrm{S}}(A_{\mathrm{P}}+1) , \qquad (4.15)$$

which is utilized as a parameter representing the resolution.[8] It should be noted that this value of x_{h} is independent of the wavelength of the incident

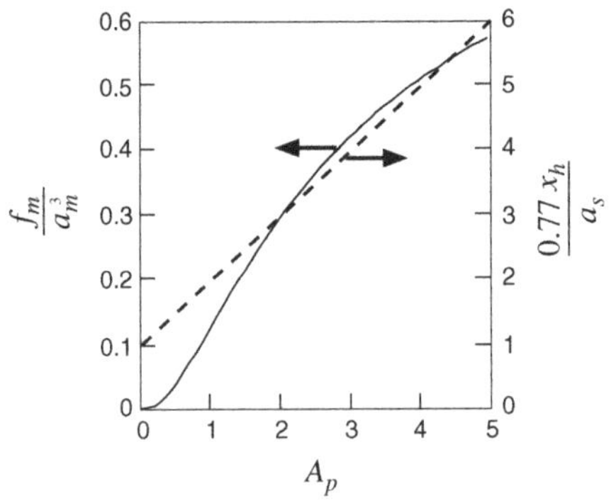

Fig. 4.6. Dependence of f_{m} and x_{h} on $A_{\mathrm{P}}(=a_{\mathrm{P}}/a_{\mathrm{S}})$

[8] An alternative parameter for representing the resolution is $x_{\mathrm{h}} - a_{\mathrm{S}}$, i.e., the difference between x_{h} and the radius a_{S} of the sphere S.

light. Further, Fig. 4.5 illustrates that the signal intensity increases dramatically when the separation R becomes smaller than $\sqrt{x_{\mathrm{h}}^2 + (a_{\mathrm{P}} + a_{\mathrm{S}})^2}$, i.e., $1.26(a_{\mathrm{P}} + a_{\mathrm{S}})$. These characteristics mean that the use of the dipole–dipole interaction can achieve high resolution with a view to measuring, fabricating, and manipulating sub-wavelength-sized object.

The value of x_{h} in (4.15) increases with increasing a_{P}, as shown in Fig. 4.6. This means that one must use a smaller sphere P to realize higher resolution, and this provides the basis for satisfying the first requirement on the sphere P presented in Sect. 3.2. For $a_{\mathrm{P}} \to 0$, the value of x_{h} becomes $0.77a_{\mathrm{S}}$, which is approximately equal to a_{S}. It also provides the basis for ensuring that the spatial distribution of the optical near field is the same as the radius of the sphere S.

In the above discussion, it was assumed that the probe touches the sphere S at $x = 0$ during the scanning (see Fig. 4.4). Next, let us assume a different case in which the sphere P is kept at a distance z from the top surface of the sphere S during scanning. In this case, the separation R is expressed as $\sqrt{x^2 + (a_{\mathrm{P}} + a_{\mathrm{S}} + z)^2}$. Also, f_{m} is $(a_{\mathrm{P}}^3 a_{\mathrm{S}}^3)/(a_{\mathrm{P}} + a_{\mathrm{S}} + z)^3$, which is less than the value in (4.14). On the other hand, x_{h} is given by $0.77(a_{\mathrm{P}} + a_{\mathrm{S}} + z)$, which is larger than the value in (4.15). This reveals that an increase in the distance z between the spheres P and S deteriorates the sensitivity and resolution of the measurement.

Figures 4.7a and b show an example of the deterioration in resolution. In this figure, flagellar filaments of salmonella bacteria fixed on a glass substrate, whose diameters are about 25 nm, are imaged by the collection mode of the near-field optical microscope [4.3]. The fiber probe used has the same

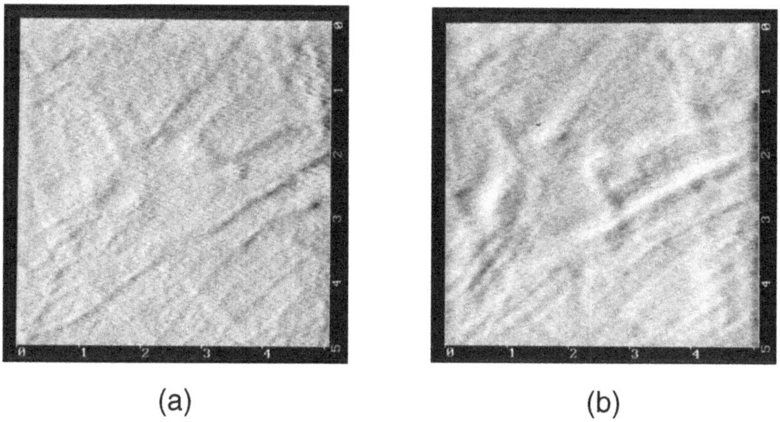

Fig. 4.7. Flagellar filaments of salmonella bacteria on a glass plate, observed by near-field optical microscope in collection mode. The foot radius a_{f} and tip radius a of the fiber probe were 15 nm and 1.5 nm, respectively. The incident light was s-polarized. Sample–probe separations z for (**a**) and (**b**) were 15 nm and 65 nm, respectively. Image size 5 μm × 5 μm

profile as shown in Fig. 2.8, where the tip radius a and foot diameter a_{f} are 1.5 nm and 15 nm, respectively. Although the images shown in Figs. 4.7a and b were taken at the same positions on the substrate, the distance z between the probe and sample were different, viz., 15 nm in Fig. 4.7a and 65 nm in Fig. 4.7b. Many flagellar filaments are observed in both figures. It is clear that the widths of the flagellar filaments in Fig. 4.7b are broader than those in Fig. 4.7a. This results from the deterioration in resolution with distance z.

In addition, it should be noted that the black strings represent the flagellar filaments, i.e., the images in these figures are reversed. We will discuss this and other features of the images in Sect. 4.2.4.

4.2.3 Contrast to Background Light

In this section, we discuss the contribution of the first term $(\alpha_{\mathrm{P}} + \alpha_{\mathrm{S}})^2 |\boldsymbol{E}_0|^2$ in (4.11). This term represents the scattered light 1, which contributes to the background signal to the scattered light 2 [the second term $4\Delta\alpha(\alpha_{\mathrm{P}} + \alpha_{\mathrm{S}})|\boldsymbol{E}_0|^2$ in (4.11)]. The contrast C of the image is defined as the ratio of the intensities of the scattered light 2 to that of 1, i.e., the ratio of the second and first terms, which yields $C = 4\Delta\alpha/(\alpha_{\mathrm{P}} + \alpha_{\mathrm{S}})$. With the help of (4.9a) and (4.9b), it is expressed as

$$\begin{aligned} C &= 4\frac{1}{2\pi\varepsilon_0 R^3}\frac{\alpha_{\mathrm{P}}\alpha_{\mathrm{S}}}{\alpha_{\mathrm{P}} + \alpha_{\mathrm{S}}} \\ &= \frac{2}{\pi\varepsilon_0} g_{\mathrm{P}} g_{\mathrm{S}} \frac{1}{\left[x^2 + (a_{\mathrm{P}} + a_{\mathrm{S}})^2\right]^{3/2}} \frac{a_{\mathrm{P}}^3 a_{\mathrm{S}}^3}{g_{\mathrm{P}} a_{\mathrm{P}}^3 + g_{\mathrm{S}} a_{\mathrm{S}}^3} , \end{aligned} \tag{4.16}$$

where the distance z is fixed at 0 for simplicity. From this equation, one finds that the contrast C has maximum value C_{m} at $x = 0$, given by

$$C_{\mathrm{m}} = \frac{2}{\pi\varepsilon_0} g_{\mathrm{P}} g_{\mathrm{S}} \frac{1}{(a_{\mathrm{P}} + a_{\mathrm{S}})^3} \frac{a_{\mathrm{P}}^3 a_{\mathrm{S}}^3}{g_{\mathrm{P}} a_{\mathrm{P}}^3 + g_{\mathrm{S}} a_{\mathrm{S}}^3} . \tag{4.17}$$

In order to evaluate C_{m}, this equation is transformed to

$$C_{\mathrm{m}} = \frac{2 g_{\mathrm{S}}}{\pi\varepsilon_0} F(G_{\mathrm{P}}, A_{\mathrm{P}}) , \tag{4.18a}$$

where $G_{\mathrm{P}} = g_{\mathrm{P}}/g_{\mathrm{S}}$ and $A_{\mathrm{P}} = a_{\mathrm{P}}/a_{\mathrm{S}}$. The function $F(G_{\mathrm{P}}, A_{\mathrm{P}})$ is defined by

$$F(G_{\mathrm{P}}, A_{\mathrm{P}}) = \frac{G_{\mathrm{P}} A_{\mathrm{P}}^3}{(A_{\mathrm{P}} + 1)^3 (G_{\mathrm{P}} A_{\mathrm{P}}^3 + 1)} , \tag{4.18b}$$

which has the maximum value $G_{\mathrm{P}}/(1 + G_{\mathrm{P}}^{1/4})^4$ at $A_{\mathrm{P}} = G_{\mathrm{P}}^{-1/4}$. Figure 4.8 depicts the profile of $F(G_{\mathrm{P}}, A_{\mathrm{P}})$ in the case $G_{\mathrm{P}} = 1$ for simplicity. For $a_{\mathrm{P}} = a_{\mathrm{S}}$, $F(G_{\mathrm{P}}, A_{\mathrm{P}})$ has maximum value 1/16. This means that the highest contrast is obtained when $a_{\mathrm{P}} = a_{\mathrm{S}}$. This phenomenon is called size-dependent resonance.

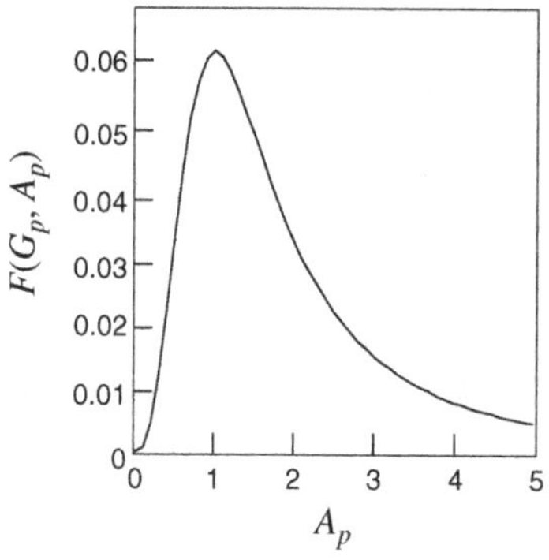

Fig. 4.8. Dependence of $F(G_{\mathrm{P}}, A_{\mathrm{P}})$ on A_{P} for $G_{\mathrm{P}} = 1$

Section 4.2.2 showed that an increase in a_{P} improves signal intensity, while deteriorating resolution. On the other hand, in this section we explained that the contrast has maximum value at $a_{\mathrm{P}} = G_{\mathrm{P}}^{-1/4} a_{\mathrm{S}}$. Therefore, we can conclude that the optimum radius of the sphere P is $G_{\mathrm{P}}^{-1/4} a_{\mathrm{S}}$ for high signal intensity, high resolution, and high contrast. This is the first requirement described in Sect. 3.2.

The contrast C_{m} has maximum value C_{mm} at $a_{\mathrm{P}} = G_{\mathrm{P}}^{-1/4} a_{\mathrm{S}}$, which is expressed as $(2g_{\mathrm{S}}/\pi\varepsilon_0)G_{\mathrm{P}}/(1 + G_{\mathrm{P}}^{1/4})^4$. In other words, when the dielectric spherical probe is used, one cannot get higher contrast than this value. In order to increase the value of C_{mm}, one solution is to increase the value of g_{P}, because C_{mm} increases with increasing G_{P}. For this purpose, we will suggest a method in Sect. 4.3.3. For $G_{\mathrm{P}} \to \infty$, C_{mm} tends to $2g_{\mathrm{S}}/\pi\varepsilon_0$. Section 4.3 suggests that one can get the value of C_{mm} larger than $2g_{\mathrm{S}}/\pi\varepsilon_0$ by using a non-spherical probe.

4.2.4 Dependence on Incident Light Polarization

This section describes the dependence of image characteristics on the polarization of the incident light in order to explain the image reversal shown in Fig. 4.7 [4.4, 4.5]. Figure 4.7 shows images measured by near-field optical microscope in collection mode. The incident light arrives at an oblique incidence angle from the rear surface of the substrate (see Fig. 3.3a). In this situation, the incident light can have two polarization states, as shown in Figs. 4.9a and b. They are called *s*- and *p*-polarized light. Their electric field vectors are oriented normal and parallel to the xz plane, respectively.[9] The characteristics of the images taken for these polarizations are discussed in the following.

[9] See Sect. 6.2.2 for further discussion of the dependence on the incident light polarization

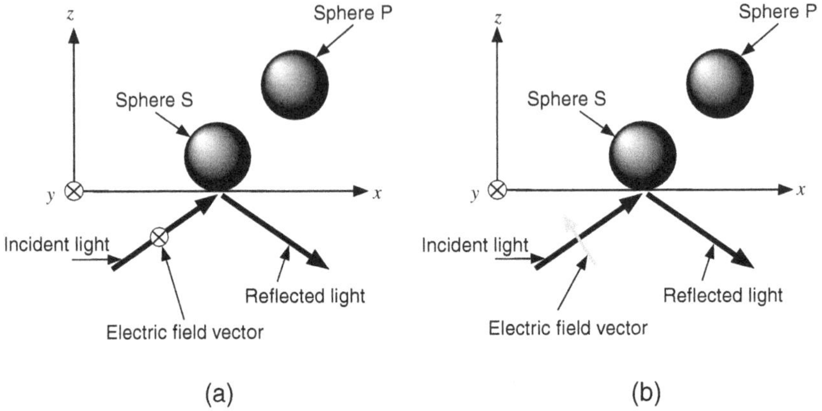

Fig. 4.9. Schematic diagram of spheres S and P for measurement with the near-field optical microscope in collection mode. Incident light is s- and p-polarized in (**a**) and (**b**), respectively

S-Polarized Light

First, assume that the sphere P is above the sphere S. Since the electric field vector of the s-polarized incident light is oriented along the y-axis as shown in Fig. 4.10, the electric dipole moments $\boldsymbol{p}_\mathrm{S}$ and $\boldsymbol{p}_\mathrm{P}$ are also oriented along the y-axis. In the figure, one can see the electric lines of force, i.e., the electric fields which are generated by $\boldsymbol{p}_\mathrm{S}$ and $\boldsymbol{p}_\mathrm{P}$. (These electric lines of force represent the near-field component of the electric field generated from the oscillating electric dipole moment. The illustration is based on the results of Sect. A.1.2 and Fig. A.1c in Appendix A.) The electric dipole moments $\Delta\boldsymbol{p}_\mathrm{S}$ and $\Delta\boldsymbol{p}_\mathrm{P}$ are generated by these electric fields. (The orientation of $\Delta\boldsymbol{p}_\mathrm{P}$ is determined by the direction of electric lines of force generated by $\boldsymbol{p}_\mathrm{S}$. Thus, one can find the orientation of $\Delta\boldsymbol{p}_\mathrm{P}$ by examining the directions of these electric lines of force at the center of the sphere P. The orientation of $\Delta\boldsymbol{p}_\mathrm{S}$ can also be found in a similar manner.) However, the orientations of $\Delta\boldsymbol{p}_\mathrm{S}$ and $\Delta\boldsymbol{p}_\mathrm{P}$ are opposite to $\boldsymbol{p}_\mathrm{S}$ and $\boldsymbol{p}_\mathrm{P}$, because they are generated by the electric field of the counterpart electric dipole moment, i.e., $\boldsymbol{p}_\mathrm{P}$ and $\boldsymbol{p}_\mathrm{S}$, respectively. Therefore, the total electric dipole moment in each sphere will be decreased by the opposite dipole moment generated by the counterpart.

Next assume that the sphere P is slightly displaced from the top of the sphere S. In this case, the orientations of $\boldsymbol{p}_\mathrm{P}$ and $\Delta\boldsymbol{p}_\mathrm{P}$ are not perfectly opposite to each other. Similarly, the orientations of $\boldsymbol{p}_\mathrm{S}$ and $\Delta\boldsymbol{p}_\mathrm{S}$ are not perfectly opposite. This means that when the sphere P is located above the sphere S, the magnitudes of the total electric dipole moments in each sphere will take the minimum value and the intensity of the scattered light 2 will be decreased. This is the reason why the inverted image was obtained in Fig. 4.7.

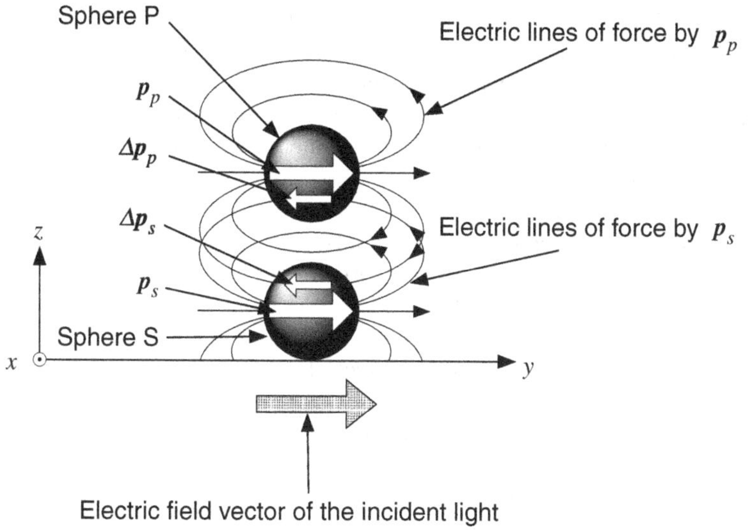

Fig. 4.10. Schematic diagram showing p_P, p_S, Δp_P, and Δp_S in the case of *s*-polarized incident light. Aspects of electric lines of force generated by p_P and p_S are also shown

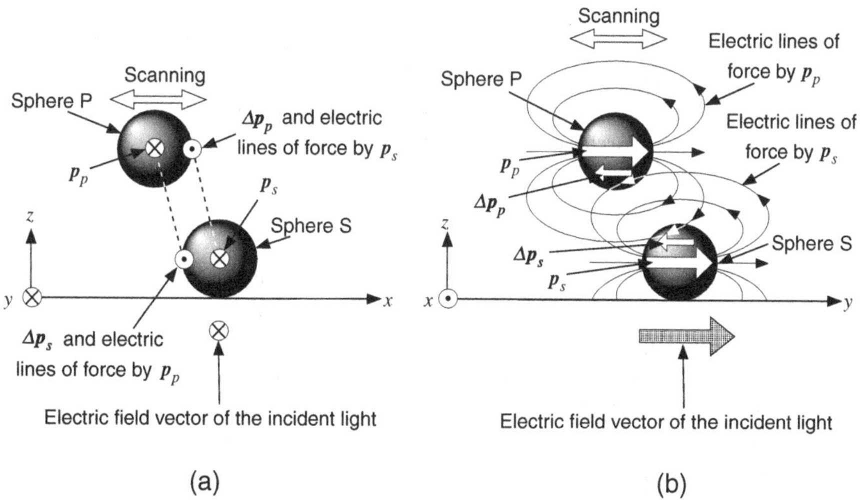

Fig. 4.11. Schematic diagrams illustrating how image characteristics depend on the scanning direction of sphere P: (**a**) along the *x*-axis and (**b**) along the *y*-axis

In the following, we will discuss the dependence of image characteristics on the scanning direction of the sphere P. Figure 4.11a shows that the sphere P is scanned along the *x*-axis. The electric dipole moment p_P and the electric field generated by it are oriented along the *y*-axis, whereas they are oriented along the tangential direction on the surface of the sphere S. This means that the electric dipole moment $\Delta \boldsymbol{p}_S$ is also oriented in the tangential direction. Therefore, when the sphere P is scanned along the x-axis, the value of $\Delta \boldsymbol{p}_S$ is insensitive to the discontinuity in the dielectric constant on the surface of the sphere S. In contrast, when the sphere P is scanned along the y-axis as shown in Fig. 4.11b, the electric dipole moment $\Delta \boldsymbol{p}_S$ induced by the electric field from $\boldsymbol{p}_P$ is oriented in the normal direction to the surface of the sphere S. Consequently, when the sphere P is scanned along the y-axis, the value of $\Delta \boldsymbol{p}_S$ is sensitively affected by the discontinuity in the dielectric constant at the surface of the sphere S. In other words, when the sphere P is at the edge of the sphere S, the scattered light intensity becomes higher than when it is above the sphere S. This phenomenon is called the edge effect. It depends on the scanning direction. It is this effect that produces the bright line seen along the flagellar filaments in Fig. 4.7.

Figures 4.12a and b show the image of a biological specimen called a microtubule, obtained from a fresh pig brain [4.4], as a clearer example revealing image inversion and the edge effect shown in Fig. 4.7. The electron microscope observation revealed that its diameter was 30 nm. Figure 4.12a is the topographical image measured by the shear-force microscope (see Sect. 3.3.3). The white line and small white spot in the figure represent the microtubule and a protein aggregate, respectively. Figure 4.12b shows the region inside the square magnified and imaged using a collection-mode near-field optical microscope. In this figure, the images of the microtuble and the protein aggregate are dark, due to image inversion as illustrated in Fig. 4.10. Further, Fig. 4.12c shows the cross-sectional profile of the light intensity along the two white arrows in Fig. 4.12b. It clearly shows that the intensity is higher at the edges, which is due to the edge effect described above.

P-Polarized Light

Assume that the sphere P is above the sphere S. In contrast to the s-polarized case, the electric field vector of the incident light is oriented along the z-axis, and the electric dipole moments $\boldsymbol{p}_S$ and $\boldsymbol{p}_P$ are also along the z-axis (see Fig. 4.13). By examining the directions of the electric lines of force of the electric field generated by $\boldsymbol{p}_S$ and $\boldsymbol{p}_P$, it can be observed that $\Delta \boldsymbol{p}_P$ and $\Delta \boldsymbol{p}_S$ induced by the electric fields in the spheres P and S are also oriented along the z-axis. Hence, $\boldsymbol{p}_S$, $\boldsymbol{p}_P$, $\Delta \boldsymbol{p}_S$, and $\Delta \boldsymbol{p}_P$ are all parallel and the scattered light intensity reaches its maximum when the sphere P is above the sphere S. The image is not reversed in this case, in contrast to the situation with s-polarized light.

Since both $\Delta \boldsymbol{p}_S$ and $\Delta \boldsymbol{p}_P$ are oriented along the z-axis, the electric lines of force generated by these electric dipole moments have bow-tie-shaped profiles spreading to the left and right of the spheres S and P. This means that the generated scattered light propagates along the x- and y-axes.[10] The scattered

[10] See Fig. A.3 in Sect. A.1.2 in Appendix A

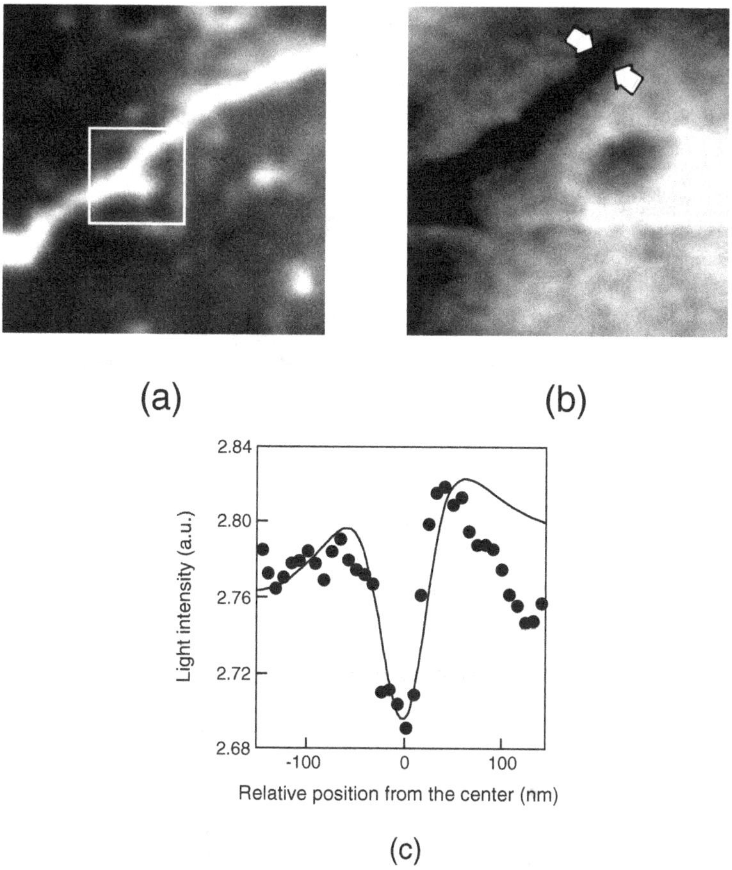

Fig. 4.12. Microtubule from a fresh pig brain on a glass plate [4.4]. (**a**) Topographical image measured by a shear-force microscope. The *small white circle* represents a protein aggregate. Image size 2 µm × 2 µm. (**b**) Magnified image of the *white square* in the center of (**a**) measured by near-field optical microscope in collection mode. Image size 550 nm × 550 nm. (**c**) Cross-sectional profile of the light intensity along the *two white arrows* in (**b**). *Black circles* and *solid curves* represent experimental and theoretical values, respectively

light intensity measured thus reaches its minimum when the sphere P is above the sphere S. The highest intensity is measured when the sphere P is displaced from above the sphere S. This phenomenon is known as the polarization-dependent edge effect.

In order to demonstrate this effect, we show the image of flagellar filaments of salmonella bacteria in Fig. 4.14. Experimental conditions are the same as in Fig. 4.7a, except that the incident light is *p*-polarized. The images in Fig. 4.14 are not reversed. Pairs of lines caused by the polarization-dependent edge effect can be seen.

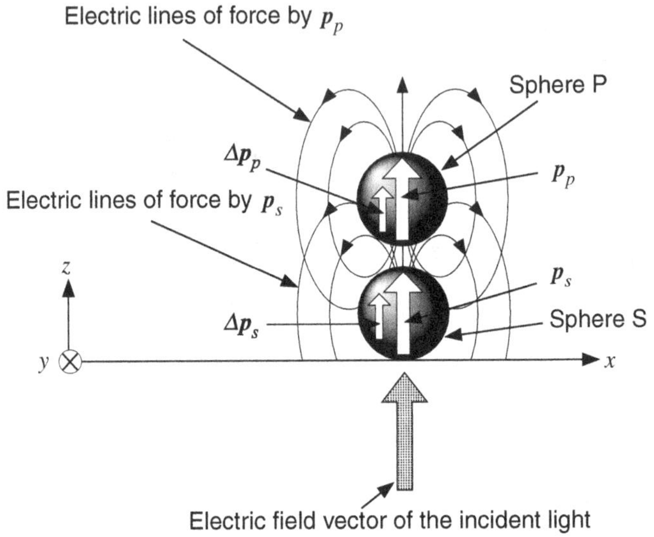

Fig. 4.13. Schematic diagrams of $\boldsymbol{p}_{\mathrm{P}}$, $\boldsymbol{p}_{\mathrm{S}}$, $\Delta\boldsymbol{p}_{\mathrm{P}}$, and $\Delta\boldsymbol{p}_{\mathrm{S}}$ in the case of p-polarized incident light. The electric lines of force generated by $\boldsymbol{p}_{\mathrm{P}}$ and $\boldsymbol{p}_{\mathrm{S}}$ are also shown

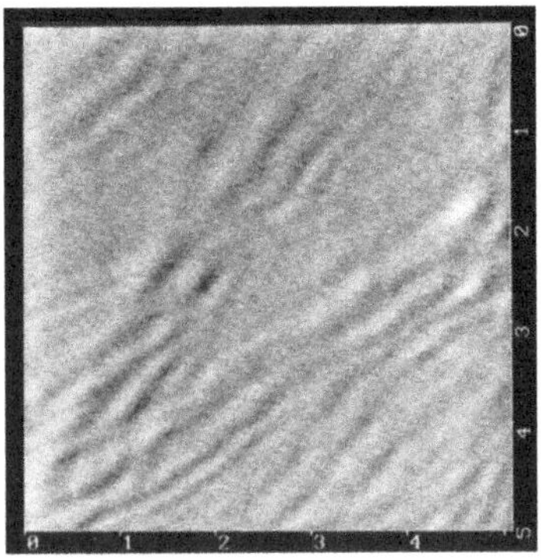

Fig. 4.14. Images of flagellar filaments of salmonella bacteria on a glass plate, observed by near-field optical microscope in collection mode. Apart from the fact that the incident light was p-polarized, the other experimental conditions are the same as in Fig. 4.7a. Image size $5\,\mu\mathrm{m} \times 5\,\mu\mathrm{m}$

4.3 Characteristics of Fiber Probes

In this section, we discuss a more practical case in which tapered fiber probes are used instead of the spherical probe.

4.3.1 Visibility and its Dependence on Cone Angle

When using a fiber probe, as shown in Figs. 2.8 and 2.10b, one has to consider the dipole–dipole interaction between the sample and the tapered part of the probe in addition to that due to the sample and probe tip. In order to consider this interaction, the profile of the fiber probe is approximated as shown in Fig. 4.15. In this figure, the fiber probe is composed of two spheres P and T, which represent the probe tip and tapered part, respectively, as shown in Fig. 3.2a. The relation between the cone angle θ and their radii a_{P}, a_{T} is

$$a_{\mathrm{T}} = \frac{1+\sin(\theta/2)}{1-\sin(\theta/2)} a_{\mathrm{P}} , \tag{4.19}$$

where $a_{\mathrm{P}} < a_{\mathrm{T}}$.

In addition to the dipole–dipole interaction between the spheres S and P, that between the spheres S and T must also be considered. Referring to (4.12), the change in polarizability of the sphere T due to this interaction is given by

$$\Delta\alpha_{\mathrm{T}} = \frac{g_{\mathrm{T}} g_{\mathrm{S}}}{2\pi\varepsilon_0} \frac{a_{\mathrm{T}}^3 a_{\mathrm{S}}^3}{R_{\mathrm{T}}^3} , \tag{4.20}$$

where $R_{\mathrm{T}} = \sqrt{x^2 + (2a_{\mathrm{P}} + a_{\mathrm{S}} + a_{\mathrm{T}})^2}$. Since the spheres P and T are made from the same material (i.e., $g_{\mathrm{T}} = g_{\mathrm{P}}$), $(a_{\mathrm{T}}^3 a_{\mathrm{S}}^3)/R_{\mathrm{T}}^3$ is transformed to $f_{\mathrm{T}}(x,\theta)$ by the definition of $f(x)$ in (4.13), where $f_{\mathrm{T}}(x,\theta)$ is given by

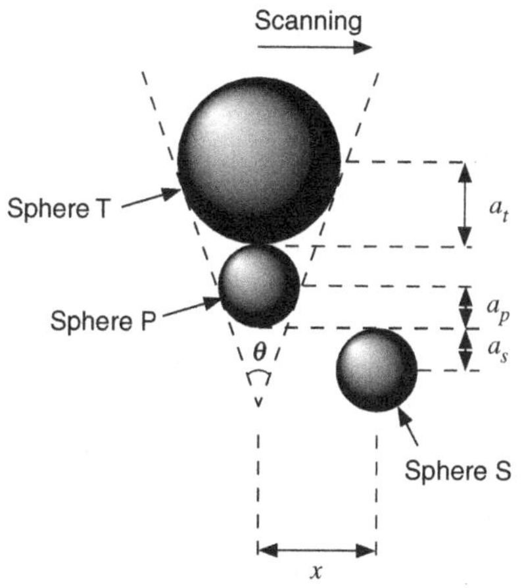

Fig. 4.15. Fiber probe approximated by two spheres S and T. θ is the cone angle

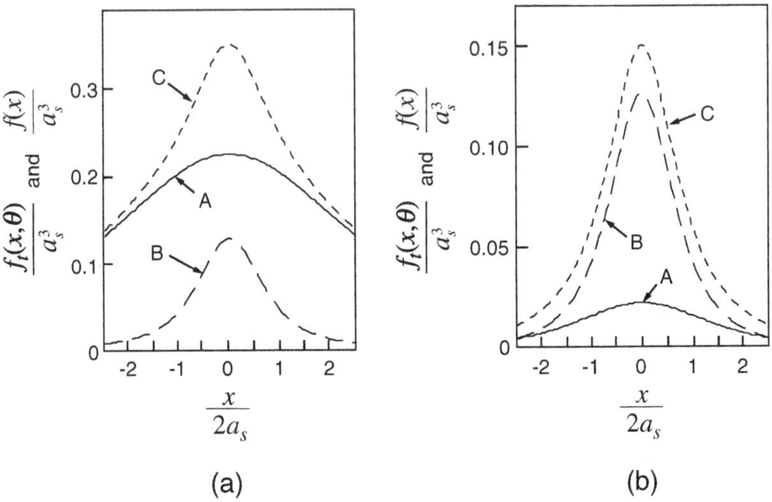

Fig. 4.16. Plots of $f_{\mathrm{T}}(x,\theta)$ and $f(x)$, where $G_{\mathrm{P}} = 1$ and $a_{\mathrm{P}} = a_{\mathrm{S}}$. Curves A and B represent the contributions from spheres T and P, respectively. Curve C is the sum of curves A and B. The values of θ in (**a**) and (**b**) are 80° and 20°, respectively

$$f_{\mathrm{T}}(x,\theta) \equiv \frac{a_{\mathrm{T}}^3 a_{\mathrm{S}}^3}{\left[x^2 + (2a_{\mathrm{P}} + a_{\mathrm{S}} + a_{\mathrm{T}})^2\right]^{3/2}} \,. \tag{4.21}$$

Curves A in Figs. 4.16a and b, based on (4.19) and (4.21), represent the profile of $f_{\mathrm{T}}(x,\theta)$ for larger and smaller cone angles θ, respectively. In the figure, the value of G_{P} ($\equiv g_{\mathrm{P}}/g_{\mathrm{S}}$) was unity, and a_{P} was equal to a_{S} in order to obtain the maximum value in the curve shown in Fig. 4.8. Curves B represent $f(x)$ of (4.13), the contribution from the sphere P. Curves C represents the total change in polarizability of the fiber probe, i.e., the sum of curves A and B, where the dipole–dipole interaction between the spheres P and T is neglected.

Comparing the curves C in Figs. 4.16a and b, one finds that the contribution from the sphere T is larger for larger cone angle θ, so that the curve C in Fig. 4.16a is wider than that in Fig. 4.16b. For a near-field optical microscope used in collection mode, the reason is as follows. In Sect. 4.2.2 it was explained that the resolution is determined by the radius a_{P} of the sphere P irrespective of the value of the cone angle θ, because the optical near field on the sample surface is scattered by P. However, in the case of the fiber probe, the contribution from scattering by the sphere T has to be taken into account. The magnitude of this contribution is larger for larger θ than that for smaller θ, because the radius of the sphere T is greater. This can be understood by plotting f_{m} as shown in Fig. 4.6. The curve C thus becomes wider for larger θ due to the larger contribution from the curve A. This is caused by the fact that light scattered by the sphere T for larger θ veils the light scattered by the sphere P.

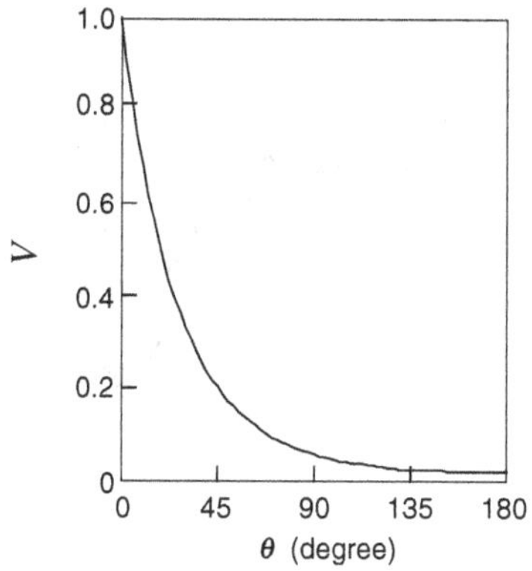

Fig. 4.17. Dependence of V on θ, where $G_{\mathrm{P}} = 1$ and $a_{\mathrm{P}} = a_{\mathrm{S}}$

The value $f_{\mathrm{m}}/f_{\mathrm{T}}(0, \theta)$ [see (4.14) and (4.21)] represents the ratio between the light intensities scattered by the sphere P and the sphere T, and this corresponds to the ratio between the heights of curves A and B. This ratio can be defined as the visibility, because it represents how clearly the high-resolution image measured by the sphere P can be seen in the low-resolution image measured by the sphere T.[11] After normalizing this ratio to unity at $\theta = 0$ (i.e., $a_{\mathrm{T}} = a_{\mathrm{P}}$), the visibility is defined by

$$V = \frac{f_{\mathrm{m}}/f_{\mathrm{T}}(0, \theta)}{f_{\mathrm{m}}/f_{\mathrm{T}}(0, 0)} = \frac{f_{\mathrm{T}}(0, 0)}{f_{\mathrm{T}}(0, \theta)} \,. \tag{4.22}$$

By using (4.14) and (4.21), it can be transformed to

$$V = \left(\frac{a_{\mathrm{P}}}{a_{\mathrm{T}}}\right)^3 \left(\frac{2a_{\mathrm{P}} + a_{\mathrm{S}} + a_{\mathrm{T}}}{3a_{\mathrm{P}} + a_{\mathrm{S}}}\right)^3 \,. \tag{4.23}$$

Further, when $G_{\mathrm{P}} = 1$ and $a_{\mathrm{P}} = a_{\mathrm{S}}$, as shown in Fig. 4.16, equation (4.19) gives

$$V = \frac{1}{8}\left[\frac{2 - \sin(\theta/2)}{1 + \sin(\theta/2)}\right]^3 \,. \tag{4.24}$$

This value decreases with increasing θ, as shown in Fig. 4.17.

Consider the case of measuring an arbitrarily shaped sample with a near-field optical microscope in collection mode. In this case, the sample shape is not always spherical, but tends to be irregular. In this situation, the optical near field on the sample surface has various spatial components depending on the sample topography. Using a tapered fiber probe, the minimum size of the spatial spread of the scattered optical near field is determined by the

[11] It should be noted that this ratio differs from the contrast C of (4.16). The contrast C is defined as the ratio with respect to the background light intensity.

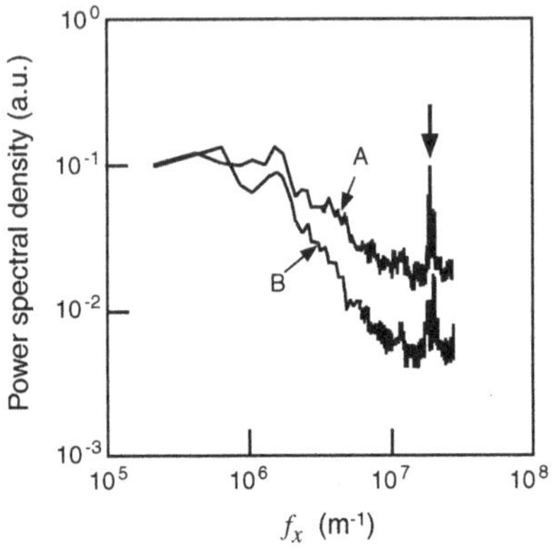

Fig. 4.18. Spatial power spectral densities of the light intensities in the images of Fig. 4.7. Curves A and B correspond to the images of Fig. 4.7a and b, respectively. The spatial Fourier frequency at the *arrow* corresponds to the reciprocal of the measured diameter of the flagellar filaments

radius a_P of the sphere P. On the other hand, the optical near field is also scattered by the sphere T, and its size depends on the size of the sphere T. Thus the intensity of the smaller-sized optical near field scattered by the sphere P is easily veiled by that of the larger-sized one scattered by the sphere T. However, if the former is sufficiently larger than the latter, the smaller-sized component can be clearly observed in the image, thus allowing high-resolution imaging.[12] The clarity is represented by the visibility V of (4.22), which is larger for smaller θ. The relation between V and θ described above is also valid for the illumination mode.

Increasing the sample–probe separation deteriorates the visibility. This is because the scattering efficiency of the optical near field by the smaller sphere P decreases more rapidly with increasing sample–probe separation than the case for the larger sphere T. The dependence of the visibility on the sample–probe separation is illustrated in Fig. 4.18. The curves A and B represent the spatial power spectral density of the light intensities shown in Figs. 4.7a and b, respectively.[13] The profiles of these curves depend on the band-pass filtering characteristics of Fig. 3.2b and the sizes of the flagellar filaments. The value of the spatial Fourier frequency f_x at the sharp peaks (indicated by an arrow) corresponds to the reciprocal of the measured diameter of the flagellar filaments. The values of the curves A and B are nearly constant in

[12] Note that this corresponds to the discussion on the band-pass filtering characteristics of Fig. 3.2b.

[13] Similar curves are shown in Fig. 3.2b. However, the horizontal axis of Fig. 4.18 is the reciprocal of that in Fig. 3.2b. It represents what is known as the spatial Fourier frequency, which corresponds to the value f_x in the solution to Problem 1.1.

the region $f_x < 10^6\,\mathrm{m}^{-1}$. However, in the region $f_x > 10^6\,\mathrm{m}^{-1}$ they decrease with increasing f_x, which means that the scattering efficiency of the smaller-sized optical near field is lower. The visibility is thus inversely proportional to the slope of these curves when $f_x > 10^6\,\mathrm{m}^{-1}$. The slope of curve B is greater than that of curve A, because the sample–probe separation in Fig. 4.7b is greater than that in Fig. 4.7a. The difference in the slopes of the curves in Fig. 4.18 thus reveals that the visibility decreases with increasing sample–probe separation.

4.3.2 Effect of Coating an Opaque Film

The third requirement in Sect. 3.2 can be satisfied by coating the foot of the tapered part of the fiber probe with an opaque film. This is because the opaque film coating of the sphere T can screen the transmission and emission of light propagating into the fiber probe in the case of collection mode and from the fiber probe in the case of illumination mode, respectively.[14] Thus, the contribution of the scattered light 1 in Fig. 2.1 can be reduced, resulting in increased contrast, as discussed in Sect. 4.2.3.

Coating with an opaque film provides further advantages. For example, the visibility can be increased by reducing the contribution of the tapered part of the fiber probe. This can be explained by replacing $f_{\mathrm{T}}(x,\theta)$ of (4.21) by $\kappa f_{\mathrm{T}}(x,\theta)$, where $\kappa(\leq 1)$ is the quantity representing the light screening effect. Figures 4.19a and b show the contribution of the tapered part for the uncoated case ($\kappa = 1$) and coated case ($\kappa = -5\,\mathrm{dB}$), respectively, where $G_{\mathrm{P}} = 1$ and $a_{\mathrm{P}} = a_{\mathrm{S}}$. The peak of the curve A in Fig. 4.19b is lower than the same in Fig. 4.19a, revealing that the contribution of the tapered part is decreased by coating with an opaque film, i.e., the visibility is increased. As a result of this decrease, the line width of curve C in Fig. 4.19b becomes narrower than that in Fig. 4.19a.

Figure 4.20 demonstrates the decrease in visibility caused by increasing the radius a_{f} of the foot, at which the sharpened fiber core protrudes from the opaque film. The figure represents an image taken under the same experimental conditions as Fig. 4.7a, except for the increase in a_{f} to 50 nm. In this figure, the profiles of the flagellar filaments cannot be seen as clearly as in Fig. 4.7a. This is due to the increased contribution of the optical near field scattered by the tapered part.

---

[14] The second requirement in Sect. 3.2 depends on the structure of the tapered part of the fiber probe. Analyzing the transmission mode through the tapered optical waveguide, one can estimate the transmission efficiency of the scattered light from the probe tip to the foot of the fiber probe for the collection mode. The same for the incident light from the foot of the fiber probe to the probe tip can also be estimated for the illumination mode. This mode analysis can be carried out using conventional waveguide theory.

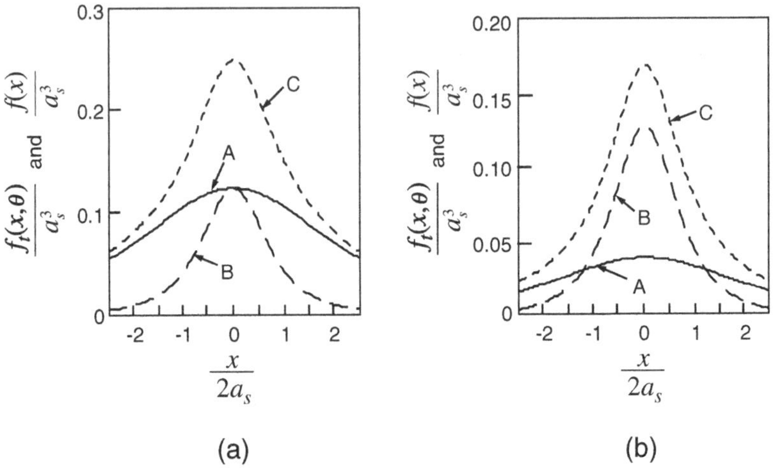

Fig. 4.19. Plots of $f_T(x, \theta)$ and $f(x)$, where $G_P = 1$ and $a_P = a_S$. The value of θ is fixed to satisfy the relation $a_T = 3a_P$. (**a**) Uncoated probe. (**b**) Probe coated with an opaque film with $\kappa = -5\,\text{dB}$. Curves A and B represent the contributions from spheres T and P, respectively. Curve C is the sum of curves A and B

Fig. 4.20. Images of the flagellar filaments of salmonella bacteria on a glass plate, observed by a near-field optical microscope in collection mode. Apart from the fact that the foot radius a_f of the fiber probe has been increased to 50 nm, the other experimental conditions are the same as in Fig. 4.7a. Image size 5 µm × 5 µm

4.3.3 Sensitivity

The efficiency of detection and generation of an optical near field can be increased by increasing the value of g_P, because $\Delta\alpha$ of (4.12) is proportional to g_P. The sensitivity of the measurement is thereby increased. Further, the maximum value C_{mm} of the contrast increases with increasing g_P, as discussed in Sect. 4.2.3. Equation (4.9c) shows that the value of g_P depends on the dielectric constant ε_P of the probe material. Since ε_P is proportional to the square of the refractive index n_P, g_P is expressed as

$$g_P = 4\pi\varepsilon_0 \frac{n_P^2 - 1}{n_P^2 + 2} , \tag{4.25}$$

where the relation $\varepsilon_P/\varepsilon_0 = n_P^2$ has been used. For $n_P^2 = -2$, the equation yields $g_P = -\infty$ and g_P increases with increasing n_P^2. Here, in order to increase the absolute value of g_P, we suggest two available methods:

- Using a material with large refractive index to make the probe. A silicon is an advantageous probe material because its refractive index can be as high as 3.4, much higher than that of a glass (≈ 1.5). An example of a probe made of silicon is shown in Fig. 3.15.
- Using a metal when making the probe. It is advantageous to coat the probe tip with a metallic film because the value of n_P^2 for a metal can be negative, e.g., -2 (see Appendix B). An example of such a fiber probe is shown in Fig. 3.1e. Its tip is coated with a metal, and so is its foot, in order to fulfill the third requirement of Sect. 3.2.

Problems

Problem 4.1

Derive the electric field at the position r generated by a static electric dipole moment $\boldsymbol{p}$. Confirm that it is equal to the one obtained by substituting $k = 0$ into (A.28) of Appendix A, viz.,

$$\boldsymbol{E} = \frac{1}{4\pi\varepsilon_0}\left[k^2(\boldsymbol{n}\times\boldsymbol{p})\times\boldsymbol{n}\left(\frac{1}{r}\right) + [3\boldsymbol{n}(\boldsymbol{n}\cdot\boldsymbol{p}) - \boldsymbol{p}]\left(-\frac{ik}{r^2} + \frac{1}{r^3}\right)e^{ikr}\right] , \tag{4.26}$$

where $\boldsymbol{n} = \boldsymbol{r}/|\boldsymbol{r}| = (x/r, y/r, z/r)$.

Problem 4.2

Derive the electric potential at $\boldsymbol{r}$ generated by a static electric dipole moment $\boldsymbol{p}$.

Problem 4.3

Prove that the polarizability of a sphere with radius a and refractive index ε is given by

$$\alpha = 4\pi\varepsilon_0\left(\frac{\varepsilon - \varepsilon_0}{\varepsilon + 2\varepsilon_0}\right)a^3 ,$$

where it is assumed that the sphere is installed in vacuum.

Problem 4.4

Assuming that the sphere P of Fig. 4.4 is scanned over the sphere S at constant elevation z, derive the maximum value of z to obtain a half width at half maximum x_h less than twice that of (4.15).

5 Electrodynamics of Oscillating Electric Dipoles

Emission spectroscopy of nanometric material systems is an important application of the optical near field (see Sect. 3.3.2), in which emission spectra are measured by a probe brought close enough to the light-emitting object. The present chapter aims to investigate the basic spectral properties of a single atom or molecule as the nanometric material system. If a conducting or dielectric probe approaches the nanomaterial, the emission properties of the atom or molecule are substantially modified. This phenomenon is discussed by treating the atom or molecule as an oscillating electric dipole moment. After presenting the basic concepts in Sect. 5.1, Sect. 5.2 gives an analytical method in which the probe tip is approximated as a planar mirror. The results of a quantum mechanical approach are described in Sect. 5.3.

5.1 Oscillating Electric Dipoles in Free Space or in a Cavity

In Sects. 5.1 and 5.2, we employ the classical model of an atom [5.1] in which an electron in the atom is considered as a classical particle with mass m and charge $-e$. In order to discuss the emission from the atom, we assume that the electron is oscillating with small displacement $\boldsymbol{a} = \boldsymbol{a}_0 \mathrm{e}^{-\mathrm{i}\omega_0 t}$, where ω_0 is the angular oscillation frequency. In this notation, it is understood that the physical displacement is the real part of the complex one. The corresponding electric dipole moment is $\boldsymbol{p}\mathrm{e}^{-\mathrm{i}\omega_0 t}$ with $\boldsymbol{p} = e\boldsymbol{a}_0$.

5.1.1 Oscillating Electric Dipole in Free Space

When the oscillating electric dipole moment $\boldsymbol{p}\mathrm{e}^{-\mathrm{i}\omega_0 t}$ (simply called the dipole from now on) is located in free space at position $\boldsymbol{x}'$, the resultant electric field at position $\boldsymbol{x}$ is

$$\boldsymbol{E}_0(\boldsymbol{r}, t) = \boldsymbol{E}_0(\boldsymbol{r})\mathrm{e}^{-\mathrm{i}\omega_0 t} , \tag{5.1}$$

where $\boldsymbol{r} = \boldsymbol{x} - \boldsymbol{x}'$. The exact expression for $\boldsymbol{E}_0(\boldsymbol{r})$ is given by (A.28) in Appendix A, which is

$$\boldsymbol{E}_0(\boldsymbol{r}) = \frac{1}{4\pi\varepsilon_0}\left[k^2(\boldsymbol{n}\times\boldsymbol{p})\times\boldsymbol{n}\left(\frac{1}{r}\right) + [3\boldsymbol{n}(\boldsymbol{n}\cdot\boldsymbol{p})-\boldsymbol{p}]\left(-\frac{\mathrm{i}k}{r^2}+\frac{1}{r^3}\right)\right]\mathrm{e}^{\mathrm{i}kr}\,, \tag{5.2}$$

where $\boldsymbol{n} = \boldsymbol{r}/|\boldsymbol{r}|$ is used. It is useful to write down the electric field for the two special cases when $\boldsymbol{p}$ is parallel to $\boldsymbol{r}$ ($\boldsymbol{n} \parallel \boldsymbol{p}$):

$$\boldsymbol{E}_0(\boldsymbol{r}) = \frac{pk^3}{2\pi\varepsilon_0}\boldsymbol{n}\left[-\frac{\mathrm{i}}{(kr)^2}+\frac{1}{(kr)^3}\right]\mathrm{e}^{\mathrm{i}kr}\,, \tag{5.3}$$

and when $\boldsymbol{p}$ is perpendicular to $\boldsymbol{r}$ ($\boldsymbol{n} \perp \boldsymbol{p}$):

$$\boldsymbol{E}_0(\boldsymbol{r}) = \frac{k^3}{4\pi\epsilon_0}\boldsymbol{p}\left[\frac{1}{kr}+\frac{\mathrm{i}}{(kr)^2}-\frac{1}{(kr)^3}\right]\mathrm{e}^{\mathrm{i}kr}\,, \tag{5.4}$$

with $p = |\boldsymbol{p}|$.

The average power radiated from the dipole into free space is obtained by integrating the flux of the far field over the surface of a large sphere, and the result is given as (see (A.34) in Appendix A)

$$P_0 = \frac{p^2k^3\omega_0}{12\pi\varepsilon_0}\,. \tag{5.5}$$

The power dissipated by the motion of the charge amounts exactly to the radiated power calculated in this equation and provides an explanation for the mechanism transferring energy from the dipole to the electromagnetic field. The corresponding radiative damping rate γ_0 is P_0/U_0, where $U_0 = m(a_0\omega_0)^2/2$ is the oscillating energy of the dipole. It follows that

$$\gamma_0 = \frac{p^2k^3}{6\pi\varepsilon_0 m\omega_0 a_0^2}\,. \tag{5.6}$$

5.1.2 Oscillating Electric Dipole in a Cavity

In order to study the phenomena induced by bringing the probe towards the atom, assume that metallic or dielectric boundaries exist close to the dipole. These boundaries are considered as a cavity even if they do not fully enclose the atom. The lowest-order perturbation approximation will be made for a low-Q cavity, in which the dipole is not strongly modified by the presence of the cavity. In that case, the cavity field $\boldsymbol{E}_\mathrm{c}$ can be expressed as a sum of the free-space field $\boldsymbol{E}_0$ in the absence of a cavity and the reflected field $\boldsymbol{E}_\mathrm{r}$ by the charges and currents induced in the walls of the cavity, i.e., $\boldsymbol{E}_\mathrm{c} = \boldsymbol{E}_0 + \boldsymbol{E}_\mathrm{r}$. It is possible to define a linear reflected field because the latter is in turn linearly induced by the dipole. This can be done by analogy with the free-space field in (5.1):

$$\boldsymbol{E}_\mathrm{r}(\boldsymbol{r},t) = \boldsymbol{E}_\mathrm{r}(\boldsymbol{r})\mathrm{e}^{-\mathrm{i}\omega_0 t}\,. \tag{5.7}$$

The precise form of this reflected field depends on the specific space and the material of the cavity. However, it should be noted that the sources are in the walls of the cavity, so the reflected field is free of divergences except at the walls.

In this perturbative approach, the angular frequency ω_0 and damping rate γ_0 found in the free space are employed as the zeroth-order approximation. When the dipole is placed in the cavity, it is perturbed by a driving force $e\boldsymbol{E}_{\mathrm{r}}(\boldsymbol{r} = 0)$ due to the reflected field, so that the displacement $a(= |\boldsymbol{a}|)$ satisfies the equation

$$\begin{aligned} \frac{\mathrm{d}^2 a}{\mathrm{d}t^2} + \gamma_0 \frac{\mathrm{d}a}{\mathrm{d}t} + \omega_0^2 a &= \frac{1}{m a_0^2} \boldsymbol{p} \cdot \boldsymbol{E}_{\mathrm{r}}(\boldsymbol{r} = 0) a \\ &= \frac{6\pi\varepsilon_0\omega_0\gamma_0}{p^2 k^3} \boldsymbol{p} \cdot \boldsymbol{E}_{\mathrm{r}}(\boldsymbol{r} = 0) a \;, \end{aligned} \tag{5.8}$$

when we assume $\gamma_0 \ll \omega_0$. The second line of this equation was rewritten by using (5.6). Solving to first order in $\boldsymbol{E}_{\mathrm{r}}(\boldsymbol{r} = 0)$ by assuming a solution of the form $a_0 \mathrm{e}^{-\mathrm{i}\omega t - (\gamma/2)t}$, the perturbed decay rate γ and angular oscillation frequency ω are obtained as

$$\gamma = \gamma_0 + \frac{6\pi\varepsilon_0\gamma_0}{p^2 k^3} \mathrm{Im}\{\boldsymbol{p} \cdot \boldsymbol{E}_{\mathrm{r}}(\boldsymbol{r} = 0)\} \;, \tag{5.9}$$

$$\omega = \omega_0 - \frac{3\pi\varepsilon_0\gamma_0}{p^2 k^3} \mathrm{Re}\{\boldsymbol{p} \cdot \boldsymbol{E}_{\mathrm{r}}(\boldsymbol{r} = 0)\} \;. \tag{5.10}$$

The symbols Im{ } and Re{ } in these equations represent the imaginary and real parts of the quantity inside the brackets, respectively.

Suppose that the cavity is initially far away, so that the dipole is regarded as existing in free space. As it is moved towards the cavity, the angular oscillation frequency shifts and, provided the change is adiabatically slow, the action U/ω is conserved. Hence, the final energy is $U = \omega U_0/\omega_0$ or, using (5.10) and $U_0 = m(a_0\omega_0)^2/2$, we obtain

$$U = U_0 - \frac{m a_0^2 \omega_0}{2} \frac{3\pi\varepsilon_0\gamma_0}{p^2 k^3} \mathrm{Re}\{\boldsymbol{p} \cdot \boldsymbol{E}_{\mathrm{r}}(\boldsymbol{r} = 0)\} \;. \tag{5.11}$$

This can be rewritten in the form

$$U - U_0 = -\frac{1}{4} \mathrm{Re}\{\boldsymbol{p} \cdot \boldsymbol{E}_{\mathrm{r}}(\boldsymbol{r} = 0)\} \;, \tag{5.12}$$

by using (5.6). It expresses the cycle-averaged interaction energy of the dipole with the reflection of its own field from the cavity walls.

5.2 Oscillating Electric Dipoles in Front of a Planar Mirror

In this section, we discuss the change in spectral properties induced when the probe approaches the light-emitting material object. This change can be induced because the probe surface acts as a cavity surface for the reflected field. For simplicity, the probe surface is assumed to be an infinite perfectly conducting plane, which we shall call a mirror. The mirror image of the dipole, given in Problem 5.1, is used as the source of the reflected field $\boldsymbol{E}_\mathrm{r}$. Equations (5.9)–(5.12) can be used for this approach.

Case I: Dipole Parallel to a Perfectly Conducting Mirror Surface

Figure 5.1a shows that the dipole $\boldsymbol{p}_\mathrm{p}$ is parallel to the mirror surface. In this case, the field to the right of the mirror is most easily found by the method of images, in which the conducting boundary is replaced by a fictitious dipole $-\boldsymbol{p}_\mathrm{p}$ situated at distance z behind the mirror, where z is the separation between the physical dipole and the mirror surface. The reflected field is just the field radiated from this image dipole. Referring to (5.4), one can express the reflected field at the site of the physical dipole as

$$\boldsymbol{E}_\mathrm{rp}(\boldsymbol{r}=0) = -\frac{k^3}{4\pi\varepsilon_0}\boldsymbol{p}_\mathrm{p}\left(\frac{1}{\phi}+\frac{\mathrm{i}}{\phi^2}-\frac{1}{\phi^3}\right)\mathrm{e}^{\mathrm{i}\phi}\,, \tag{5.13a}$$

with $\phi = 2kz$. The modified decay rate, angular frequency, and energy are obtained by substituting this into (5.9)–(5.11):

$$\gamma_\mathrm{p} = \gamma_0 - \frac{3}{2}\gamma_0\left(\frac{\sin\phi}{\phi}+\frac{\cos\phi}{\phi^2}-\frac{\sin\phi}{\phi^3}\right)\,, \tag{5.13b}$$

$$\omega_\mathrm{p} = \omega_0 + \frac{3}{4}\gamma_0\left(\frac{\cos\phi}{\phi}-\frac{\sin\phi}{\phi^2}-\frac{\cos\phi}{\phi^3}\right)\,, \tag{5.13c}$$

$$U_\mathrm{p} = U_0 + \frac{p^2k^3}{16\pi\varepsilon_0}\left(\frac{\cos\phi}{\phi}-\frac{\sin\phi}{\phi^2}-\frac{\cos\phi}{\phi^3}\right)\,, \tag{5.13d}$$

where p in (5.13d) represents $|\boldsymbol{p}_\mathrm{p}|$.

Case II: Dipole Normal to a Perfectly Conducting Mirror Surface

For the electric dipole $\boldsymbol{p}_\mathrm{s}$ normal to the mirror, as shown in Fig. 5.1b, the image dipole is $+\boldsymbol{p}_\mathrm{s}$. The reflected field is given by (5.3) and is expressed as

$$\boldsymbol{E}_\mathrm{rs}(\boldsymbol{r}) = \frac{k^3}{2\pi\varepsilon_0}\boldsymbol{p}_\mathrm{s}\left(-\frac{\mathrm{i}}{\phi^2}+\frac{1}{\phi^3}\right)\mathrm{e}^{\mathrm{i}\phi}\,. \tag{5.14a}$$

Thus, using (5.9)–(5.12),

$$\gamma_\mathrm{s} = \gamma_0 - 3\gamma_0\left(\frac{\cos\phi}{\phi^2}-\frac{\sin\phi}{\phi^3}\right)\,, \tag{5.14b}$$

$$\omega_\mathrm{s} = \omega_0 - \frac{3}{2}\gamma_0\left(\frac{\sin\phi}{\phi^2}+\frac{\cos\phi}{\phi^3}\right)\,, \tag{5.14c}$$

$$U_\mathrm{s} = U_0 - \frac{p^2k^3}{8\pi\varepsilon_0}\left(\frac{\sin\phi}{\phi^2}+\frac{\cos\phi}{\phi^3}\right)\,, \tag{5.14d}$$

where p in (5.14d) represents $|\boldsymbol{p}_\mathrm{s}|$.

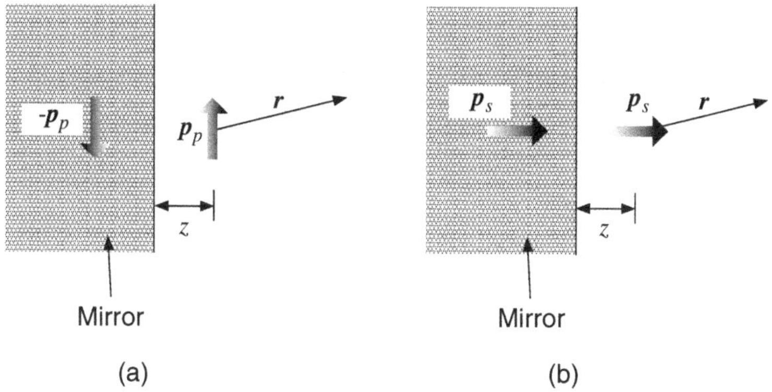

Fig. 5.1. Schematic explanation of the method of mirror images. (**a**) and (**b**) correspond to the cases when the electric dipole moment is parallel and normal to the mirror surface, respectively

Dependence of γ on Distance z

Figure 5.2 shows the radiation rates γ_{p} and γ_{s} (normalized to γ_0) as a function of distance z from the mirror (normalized to wavelength λ). These were drawn by using (5.13b) and (5.14b), respectively. When the dipole is located at many wavelengths away from the mirror, the reflected field is weak and the radiation rate is therefore close to the free-space value. As the distance from the mirror decreases, the cavity effect becomes appreciable and the radiation rate is alternatively raised and lowered as the dipole and the reflected field $\boldsymbol{E}_{\mathrm{r}}$ come in and out of phase with each other.

At short range, the radiation rate γ_{p} is suppressed while γ_{s} is twice the free-space value. This is due to the interference between $\mathrm{Im}(\boldsymbol{E}_{\mathrm{r}})$ and $\mathrm{Im}(\boldsymbol{E}_0)$, which is destructive in the first case and constructive in the second. From another point of view, the parallel dipole $\boldsymbol{p}_{\mathrm{p}}$ and its mirror image $-\boldsymbol{p}_{\mathrm{p}}$ are out-of-phase antennae. When their separation is much less than a wavelength, the far field interferes destructively. The Poynting vector vanishes, and hence no energy is radiated.

By contrast, the far field from the perpendicular dipole $\boldsymbol{p}_{\mathrm{s}}$ and its mirror image $+\boldsymbol{p}_{\mathrm{s}}$ interfere constructively when their separation is zero to produce four times the energy flux of the free dipole over the hemisphere in front of the mirror, i.e., twice the radiated power.

Dependence of ω on Distance z

The shifts in the angular oscillation frequencies ω_{p} and ω_{s} (normalized to ω_0) shown in Fig. 5.3 behave in a qualitatively similar way at long range to the shifts in the radiation rates. The angular frequencies are very strongly modified at distances within $\lambda/4$ of the surface, where the electric dipole is influenced by its image.

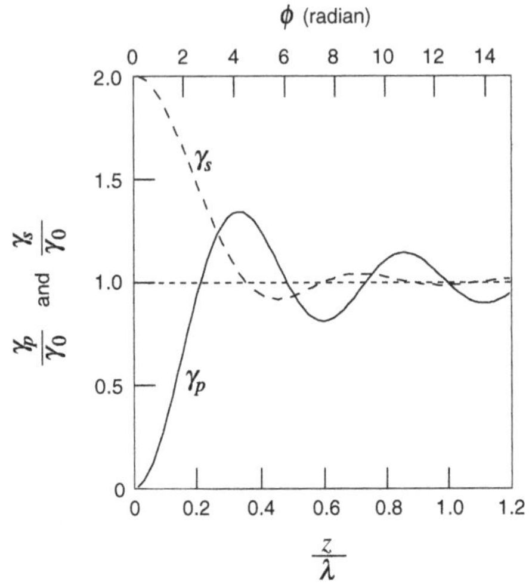

Fig. 5.2. Radiation rate as a function of distance z from the mirror. *Solid* and *broken curves* represent the values for γ_{p} and γ_{s}, respectively

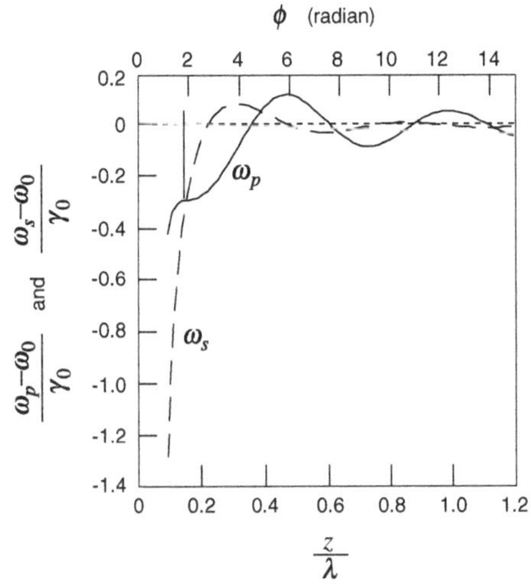

Fig. 5.3. Angular oscillation frequency as a function of distance z from the mirror. *Solid* and *broken curves* represent the values for ω_{p} and ω_{s}, respectively

Dependence of U on Distance z

At short range, the propagation delay of the reflected field is much less than an oscillation period (i.e., $\phi \ll 2\pi$), so that the interaction between the dipole and the mirror is just that of a dipole with its instantaneous image. The energy shift is

$$U_i - U_0 \propto z^{-3} \quad (i = \mathrm{p}, \mathrm{s}) \ , \tag{5.15}$$

as derived from (5.13) and (5.14), which causes a z^{-4} attractive force between the dipole and the surface. This corresponds to a classical version of the well-known van der Waals interaction between an atom and a surface.[1]

Discussions given above were for the case of a perfect conducting mirror approaching the atom or molecule. The field reflected from a real mirror has a phase shift that is not exactly 0 or π. Formally, it should be multiplied by a reflection coefficient $\zeta e^{i\delta}$, where $\zeta < 1$. This has the effect in (5.13) and (5.14) of replacing $\cos\phi$ and $\sin\phi$ by $\zeta\cos(\phi+\delta)$ and $\zeta\sin(\phi+\delta)$, respectively.

5.3 Cavity Quantum Electrodynamics of Oscillating Electric Dipoles

In order to discuss the topics of the previous sections in the context of quantum theory, one has to include the excited and ground states of the electronic energy of the atom or molecule into the theory. This is called cavity quantum electrodynamics. The main results of this theory are reviewed in this section, while the reader can refer to [5.1] for the details.

Figure 5.4 shows the calculated energy shift ΔE_g of the sodium ground state as a function of distance z from a plane mirror.[2] The decrease in ΔE_g caused by decreasing z in this figure represents the decrease in the angular frequency of light emission. This corresponds to the characteristics of the curves in Fig. 5.3. At short range ($z/\lambda \ll 1$), the magnitude of ΔE_g tends to be proportional to z^{-3}, as shown by the broken line in Fig. 5.4. It corresponds to the instantaneous van der Waals interaction. The magnitude of the shift exactly equals that of a classical dipole with the same mean square strength. At long range, the magnitude of ΔE_g tends to be proportional to z^{-4}, as shown by the dotted line in Fig. 5.4. This corresponds to the Casimir–Polder potential. The behavior shown in Fig. 5.4 bears no resemblance at all to the oscillatory shift of a classical dipole shown in Fig. 5.3. This lack of oscillation emphasizes the fact that a ground-state atom is fundamentally different from a classical dipole, in that its mean square dipole moment is nonzero and yet it cannot radiate.

By contrast, when the atom is in an excited state, the magnitude of ΔE_g tends to be proportional to z^{-3} at short range, and one finds once again the van der Waals shift. Excited states and ground states have the same behavior. At long range, however, the broadband Casimir–Polder shift is completely

[1] In the supplement to Sect. 3.3.1 it was pointed out that the magnitude of the potential energy of the interaction between the two atoms was proportional to z^{-6}. In contrast to this, an atom is replaced by a mirror for the present discussion. Since an infinite number of atoms are arranged on the mirror surface, the total potential energy is the superposition of the energies of interaction with these arranged atoms. Its magnitude is then proportional to z^{-3}, as shown by (5.15).

[2] This figure was produced by the authors by revising Fig. 10 of [5.1].

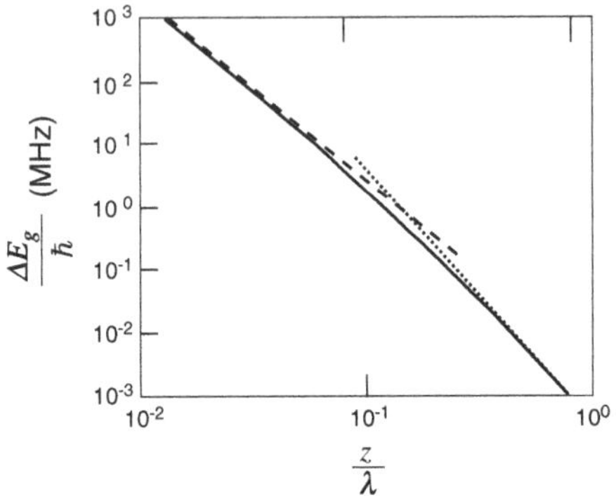

Fig. 5.4. Calculated energy shift ΔE_{g} of the sodium ground state as a function of distance z from a plane mirror. The *solid curve* represents the calculated result derived quantum mechanically. The *broken* and *dotted lines* are fitted to the *solid curves*, which are proportional to z^{-3} (van der Waals shift) and z^{-4} (Casimir–Polder shift), respectively

overwhelmed in excited states by the resonant shift. The variation of this shift with distance is identical to that of the classical dipole.

To summarize, the energy shift of an atom in front of a mirror exhibits asymptotically three physically distinct phenomena: the van der Waals shift, the Casimir–Polder shift, and the resonant radiative shift. The first phenomenon is found when the atom is close to the mirror and is a purely semi-classical effect, being due to fluctuations in the instantaneous electric dipole moment of the atom. The second phenomenon is found in ground-state atoms far from the mirror and is purely quantum electrodynamic, being due to fluctuations in the vacuum field. The third phenomenon occurs in excited atoms far from the mirror and closely resembles the interaction of a classical dipole with its reflected field.

Problems

Problem 5.1

Consider a planar dielectric with dielectric constant ε and a point charge $+q$ at position $A(a,0,0)$ in vacuum, as shown in Fig. 5.5. In order to derive the electric field in vacuum, the dielectric is replaced by a fictitious mirror image $-q'$, which is fixed at position $B(-a,0,0)$. This replacement can make the derivation easier because a boundary condition on the planar dielectric surface is removed. To derive the electric field in the dielectric, on the other hand, the charge $+q$ at position A is replaced by a fictitious charge $+q''$. Derive the values of $-q'$ and $+q''$.

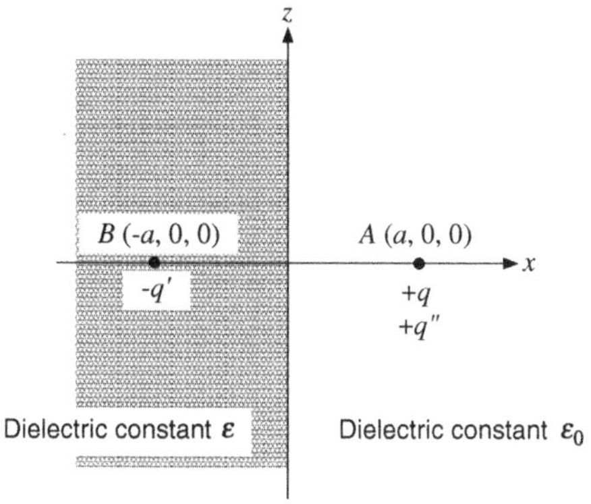

Fig. 5.5. Planar dielectric with dielectric constant ε located in $x \leq 0$. The charge $+q$ is at the position $A(a, 0, 0)$ and its fictitious mirror image $-q'$ at the position $B(-a, 0, 0)$. The charge $+q''$ at position A is also fictitious

6 Self-Consistent Method Using a Propagator

In Chap. 4, we derived the change in the polarizability of the sphere P, which was induced by the electric field from an electric dipole in the sphere S. In this derivation, the effect of multiple scattering was neglected, i.e., we neglected the changes in the polarizability of the sphere S induced by the above-mentioned change in the polarizability of the sphere P. The present chapter discusses the effect of multiple scattering for the more precise investigation of an optical near field. A propagator, i.e., the transfer function, is derived in Sect. 6.1, in order to evaluate the electric field at an arbitrary position generated by a light source at another position. The result of this derivation is applied to collection-mode near-field optical microscopy in Sect. 6.2. It should be noted that these results can be applied, not only to the two spheres S and P, but also to arbitrarily shaped material objects. However, a long computation time is required to derive quantitative results in such numerical analysis.

6.1 Propagator

This section studies the temporal and spatial characteristics of the electric field $\boldsymbol{E}$ of light generated from the induced electric dipole moments in the spheres S and P based on the formalism of classical electromagnetism. The polarization $\boldsymbol{P}$ (i.e., the vectorial sum of the electric dipole moments) is used for a more general discussion, instead of a single electric dipole moment.

6.1.1 Propagator in Free Space

By assuming a sinusoidally oscillating electric field $\boldsymbol{E}$ and the polarization $\boldsymbol{P}$ (which are represented by $\mathrm{e}^{-\mathrm{i}\omega t}$), (A.17) in Appendix A can be expressed as to

$$\nabla \times \nabla \times \boldsymbol{E}(\boldsymbol{r}) - k^2 \boldsymbol{E}(\boldsymbol{r}) = \mu_0 \omega^2 \boldsymbol{P}(\boldsymbol{r}) , \tag{6.1}$$

where we used the formula $\partial^2/\partial t^2 = -\omega^2$.[1] This equation represents the temporal and spatial characteristics of $\boldsymbol{E}$. On the other hand, $\boldsymbol{P}$ is expressed

[1] The differential operator ∇ in this equation is defined in Sect. A.1.

by using the electric susceptibility $\chi(\boldsymbol{r}, \boldsymbol{r}')$ of the material object under study, which is given by

$$\boldsymbol{P}(\boldsymbol{r}) = \int \chi(\boldsymbol{r}, \boldsymbol{r}')\delta(\boldsymbol{r} - \boldsymbol{r}')\boldsymbol{E}(\boldsymbol{r}')\mathrm{d}^3r' . \tag{6.2}$$

The integral in this equation is taken over the whole volume of the material object, whilst the delta function $\delta(\boldsymbol{r}, \boldsymbol{r}')$ is used to express the fact that $\boldsymbol{P}$ is generated locally.

Equation (6.1) shows that $\boldsymbol{E}$ is generated by $\boldsymbol{P}$, while (6.2) shows that $\boldsymbol{P}$ is induced by $\boldsymbol{E}$. Thus, one has to solve these equations in a consistent manner [6.1, 6.2], using the so-called self-consistent method. Assuming a homogeneous electric field in the material object for simplicity, the right-hand side of (6.2) is proportional to $\int \chi(\boldsymbol{r}')\mathrm{d}^3r'$ which can be replaced by the polarizability α_i. Since the polarizability for a sphere of radius r_i is expressed as

$$\alpha_i = 4\pi\varepsilon_0 \left(\frac{\varepsilon - \varepsilon_0}{\varepsilon + 2\varepsilon_0}\right) r_i^3 , \tag{6.3}$$

as given in Problem 4.3 of Chap. 4, (6.2) is transformed to $\boldsymbol{P}(\boldsymbol{r}) = \alpha_i \boldsymbol{E}(\boldsymbol{r})$, with which (6.1) is solved simultaneously.

Supplement: Green Function

The Green function $G(\boldsymbol{r}, \boldsymbol{r}')$ has been popularly used in order to obtain a solution $\phi(\boldsymbol{r})$ of the differential equation

$$(\nabla^2 + k^2)\phi(\boldsymbol{r}) = -\frac{1}{\varepsilon_0} g(\boldsymbol{r}) , \tag{6.4}$$

when $g(\boldsymbol{r})$ is a known function. It is defined by

$$G(\boldsymbol{r}, \boldsymbol{r}') = \frac{\exp(\mathrm{i}k|\boldsymbol{r} - \boldsymbol{r}'|)}{4\pi\varepsilon_0|\boldsymbol{r} - \boldsymbol{r}'|} , \tag{6.5}$$

which satisfies the equation

$$(\nabla^2 + k^2)G(\boldsymbol{r}, \boldsymbol{r}') = -\frac{1}{\varepsilon_0}\delta(\boldsymbol{r} - \boldsymbol{r}') , \tag{6.6}$$

where $\delta(\boldsymbol{r} - \boldsymbol{r}')$ is the delta function. Noting that (6.6) leads to

$$\begin{aligned}(\nabla^2 + k^2)\int G(\boldsymbol{r}, \boldsymbol{r}')g(\boldsymbol{r}')\mathrm{d}^3r' &= -\int \frac{1}{\varepsilon_0}\delta(\boldsymbol{r} - \boldsymbol{r}')g(\boldsymbol{r}')\mathrm{d}^3r' \\ &= -\frac{1}{\varepsilon_0} g(\boldsymbol{r}) , \end{aligned}\tag{6.7}$$

it is found that the solution $\phi(\boldsymbol{r})$ of (6.4) is given by

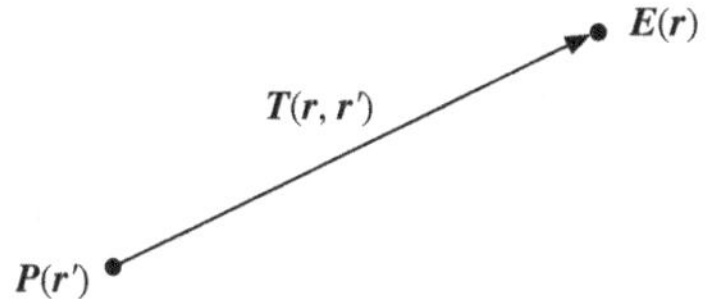

Fig. 6.1. Schematic explanation of the propagator $\boldsymbol{T}(\boldsymbol{r}, \boldsymbol{r}')$

$$\phi(\boldsymbol{r}) = \phi_0(\boldsymbol{r}) + \int G(\boldsymbol{r}, \boldsymbol{r}')g(\boldsymbol{r}')\mathrm{d}^3r' \,, \tag{6.8}$$

where the function $\phi_0(\boldsymbol{r})$ is the solution of the homogeneous differential equation, i.e., $(\nabla^2 + k^2)\phi_0(\boldsymbol{r}) = 0$. One can confirm that (6.8) is a solution of (6.4) as follows. By applying the differential operator $(\nabla^2 + k^2)$ to both sides of (6.8), one obtains

$$(\nabla^2 + k^2)\phi(\boldsymbol{r}) = (\nabla^2 + k^2)\phi_0(\boldsymbol{r}) + (\nabla^2 + k^2) \int G(\boldsymbol{r}, \boldsymbol{r}')g(\boldsymbol{r}')\mathrm{d}^3r' \,. \tag{6.9}$$

The first term $(\nabla^2 + k^2)\phi_0(\boldsymbol{r})$ in this equation is 0, and the second term is equal to $-(1/\varepsilon_0)g(\boldsymbol{r})$ by referring to (6.7). Thus, $\phi(\boldsymbol{r})$ is a solution to (6.4).

In order to derive the solution $\boldsymbol{E}(\boldsymbol{r})$ of (6.1), the Green function

$$\boldsymbol{T}(\boldsymbol{r}, \boldsymbol{r}') \equiv (k^2\boldsymbol{I} + \nabla\nabla)G(\boldsymbol{r}, \boldsymbol{r}') \tag{6.10}$$

is used. Here, $G(\boldsymbol{r}, \boldsymbol{r}')$ is defined by (6.5). The unit operator $\boldsymbol{I}$ acts by $\boldsymbol{I}\boldsymbol{x} = \boldsymbol{x}$ for any vector $\boldsymbol{x}$. Using this function and (6.10), the solution of (6.1) is expressed as

$$\boldsymbol{E}(\boldsymbol{r}) = \boldsymbol{E}_0(\boldsymbol{r}) + \int \boldsymbol{T}(\boldsymbol{r}, \boldsymbol{r}')\boldsymbol{P}(\boldsymbol{r}')\mathrm{d}^3r' \,, \tag{6.11}$$

where $\boldsymbol{E}_0(\boldsymbol{r})$ is the solution of the homogeneous differential equation. The Green function $\boldsymbol{T}(\boldsymbol{r}, \boldsymbol{r}')$ is called the propagator because it describes the relation between an electric field $\boldsymbol{E}$ at one position $\boldsymbol{r}$ and a field locally generated from a unit polarization $\boldsymbol{P}$ at the other position $\boldsymbol{r}'$, as shown in Fig. 6.1. It is a member of the family of Green functions, but it is a tensor, while $G(\boldsymbol{r}, \boldsymbol{r}')$ is a scalar function.

The propagator $\boldsymbol{T}(\boldsymbol{r}, \boldsymbol{r}')$ satisfies the equation[2]

$$\nabla \times \nabla \times \boldsymbol{T}(\boldsymbol{r}, \boldsymbol{r}') - k^2\boldsymbol{T}(\boldsymbol{r}, \boldsymbol{r}') = \mu_0\omega^2\delta(\boldsymbol{r} - \boldsymbol{r}') \,. \tag{6.12}$$

Further, it can be expressed as[3]

$$\begin{aligned} \boldsymbol{T}(\boldsymbol{r}, \boldsymbol{r}') &= \frac{1}{4\pi\varepsilon_0}\left[k^2(\boldsymbol{I} - \boldsymbol{n}\boldsymbol{n})\frac{1}{R} + (3\boldsymbol{n}\boldsymbol{n} - \boldsymbol{I})\left(-\frac{\mathrm{i}k}{R^2} + \frac{1}{R^3}\right)\right]\mathrm{e}^{\mathrm{i}kR} \\ &\equiv \left[\boldsymbol{T}_1(\boldsymbol{r}, \boldsymbol{r}') + \boldsymbol{T}_2(\boldsymbol{r}, \boldsymbol{r}') + \boldsymbol{T}_3(\boldsymbol{r}, \boldsymbol{r}')\right]\mathrm{e}^{\mathrm{i}kR} \,, \end{aligned} \tag{6.13}$$

where $\boldsymbol{n}$ is the unit vector $\boldsymbol{R}/|\boldsymbol{R}|$ with $\boldsymbol{R} = \boldsymbol{r} - \boldsymbol{r}'$ and $R = |\boldsymbol{R}|$.

[2] See Problem 6.1

[3] See Problem 6.2

The terms $\boldsymbol{T}_1(\boldsymbol{r},\boldsymbol{r}')$, $\boldsymbol{T}_2(\boldsymbol{r},\boldsymbol{r}')$, and $\boldsymbol{T}_3(\boldsymbol{r},\boldsymbol{r}')$ in this equation are proportional to R^{-1}, R^{-2}, and R^{-3}, respectively, and are given by

$$\boldsymbol{T}_1(\boldsymbol{r},\boldsymbol{r}') = \frac{1}{4\pi\varepsilon_0}k^2(\boldsymbol{I}-\boldsymbol{nn})\frac{1}{R}\,, \tag{6.14a}$$

$$\boldsymbol{T}_2(\boldsymbol{r},\boldsymbol{r}') = -\frac{1}{4\pi\varepsilon_0}(3\boldsymbol{nn}-\boldsymbol{I})\frac{ik}{R^2}\,, \tag{6.14b}$$

$$\boldsymbol{T}_3(\boldsymbol{r},\boldsymbol{r}') = \frac{1}{4\pi\varepsilon_0}(3\boldsymbol{nn}-\boldsymbol{I})\frac{1}{R^3}\,. \tag{6.14c}$$

The term $\boldsymbol{T}_3(\boldsymbol{r},\boldsymbol{r}')$ is a near-field component because its magnitude is larger than those of $\boldsymbol{T}_2(\boldsymbol{r},\boldsymbol{r}')$ and $\boldsymbol{T}_1(\boldsymbol{r},\boldsymbol{r}')$ if $kR \ll 1$. Their ratio is $(kR)^{-2} : (kR)^{-1} : 1$.

By using a Cartesian coordinate, $\boldsymbol{T}_3(\boldsymbol{r},\boldsymbol{r}')$ can also be expressed as

$$\boldsymbol{T}(\boldsymbol{r},\boldsymbol{r}') = [T_{\alpha\beta}]\,, \quad (\alpha,\beta = x,y,z)\,, \tag{6.15}$$

with

$$T_{\alpha\beta} = \frac{1}{4\pi\varepsilon_0}\left[k^2(\delta_{\alpha\beta}-n_\alpha n_\beta)\frac{1}{R} + (3n_\alpha n_\beta - \delta_{\alpha\beta})\left(-\frac{ik}{R^2}+\frac{1}{R^3}\right)\right]\mathrm{e}^{ikR}\,, \tag{6.16}$$

where $\delta_{\alpha\beta}$ is the Kronecker delta, and we denote

$$\boldsymbol{n} = (n_x,n_y,n_z) = \left(\frac{x-x'}{R},\frac{y-y'}{R},\frac{z-z'}{R}\right)\,,$$

and $R = \sqrt{(x-x')^2+(y-y')^2+(z-z')^2}$.

6.1.2 Propagator in Close Proximity to a Planar Substrate

Now that the propagator in free space has been introduced, the polarization will hereafter be replaced by a single electric dipole moment for simplicity. With this replacement, the present section describes the case in which a planar dielectric substrate with dielectric constant ε is placed in close proximity to an electric dipole moment at position $\boldsymbol{r}' = (x',y',z')$ (see Fig. 6.2). Following the solution to Problem 5.1, a mirror image dipole can be assumed at position $\boldsymbol{r}'_{\mathrm{M}} = (x'_{\mathrm{M}},y'_{\mathrm{M}},z'_{\mathrm{M}}) = (x',y',-z')$. The propagator $\boldsymbol{T}_{\mathrm{M}}(\boldsymbol{r},\boldsymbol{r}'_{\mathrm{M}})$ for this mirror image dipole is given as

$$\boldsymbol{T}_{\mathrm{M}}(\boldsymbol{r},\boldsymbol{r}'_{\mathrm{M}}) = \left(\frac{\varepsilon-\varepsilon_0}{\varepsilon+\varepsilon_0}\right)\boldsymbol{T}_3(\boldsymbol{r},\boldsymbol{r}'_{\mathrm{M}})\boldsymbol{M}\,, \tag{6.17}$$

by using (6.13). Only the near-field component $\boldsymbol{T}_3(\boldsymbol{r},\boldsymbol{r}'_{\mathrm{M}})$ is employed on the right-hand side of this equation because the substrate surface is close to the electric dipole moment. The term $(\varepsilon-\varepsilon_0)/(\varepsilon+\varepsilon_0)$ represents the magnitude of the mirror image dipole in the dielectric substrate, as given in (Q5.7) of

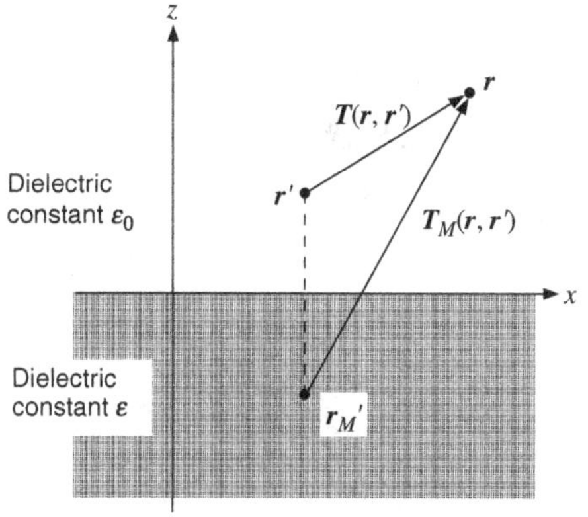

Fig. 6.2. Schematic explanation of the propagator in close proximity to the planar substrate with dielectric constant ε

the solution to Problem 5.1. The matrix $\boldsymbol{M}$ represents coordinate inversion, which is expressed as

$$\boldsymbol{M} = \begin{pmatrix} -1, & 0, & 0 \\ 0, & -1, & 0 \\ 0, & 0, & +1 \end{pmatrix} . \tag{6.18}$$

With the help of (6.15), (6.16) and (6.18), cartesian components of (6.17) are written as

$$\boldsymbol{T}_{\mathrm{M}}(\boldsymbol{r}, \boldsymbol{r}') = \frac{1}{4\pi\varepsilon_0} \frac{\varepsilon - \varepsilon_0}{\varepsilon + \varepsilon_0} \frac{1}{R^3} \mathrm{e}^{\mathrm{i}kR} \begin{pmatrix} -3n_{\mathrm{M}x}^2 + 1 & -3n_{\mathrm{M}x}n_{\mathrm{M}y} & 3n_{\mathrm{M}x}n_{\mathrm{M}z} \\ -3n_{\mathrm{M}y}n_{\mathrm{M}x} & -3n_{\mathrm{M}y}^2 + 1 & 3n_{\mathrm{M}y}n_{\mathrm{M}z} \\ -3n_{\mathrm{M}z}n_{\mathrm{M}x} & -3n_{\mathrm{M}z}n_{\mathrm{M}y} & 3n_{\mathrm{M}z}^2 - 1 \end{pmatrix} , \tag{6.19}$$

where we used the following notation

$$\begin{aligned} \boldsymbol{n}_{\mathrm{M}} &= (n_{\mathrm{M}x}, n_{\mathrm{M}y}, n_{\mathrm{M}z}) \\ &= \left(\frac{x - x'_{\mathrm{M}}}{R}, \frac{y - y'_{\mathrm{M}}}{R}, \frac{z - z'_{\mathrm{M}}}{R} \right) = \left(\frac{x - x'}{R}, \frac{y - y'}{R}, \frac{z + z'}{R} \right) . \end{aligned}$$

6.2 Application to Collection-Mode Near-Field Optical Microscopy

In this section we study collection-mode near-field optical microscopy using the self-consistent method. As explained schematically in Fig. 6.3, the fiber probe is assumed to be an ensemble of N spheres ($i = 1, \ldots, N$). On the other hand, the sample is assumed to be a two dimensional array of M spheres ($i = N+1, \ldots, N+M$) on the substrate. Further, these spheres are assumed to have polarizability α_i ($i = 1, \ldots, N+M$) at their centers r_i, where α_i is given by (6.3). Incident light impinges upon the rear surface of the substrate.

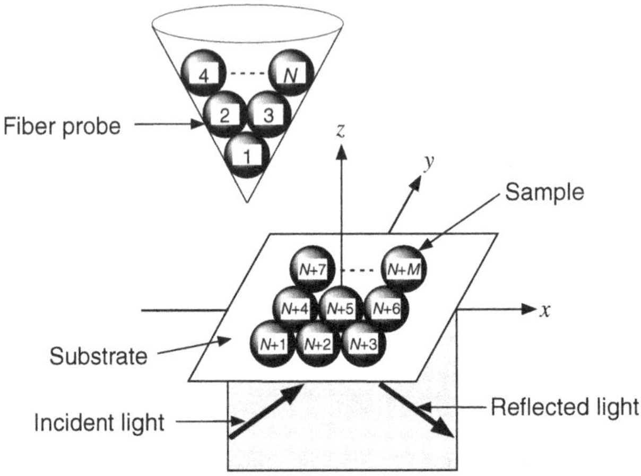

Fig. 6.3. Calculation model for collection-mode near-field optical microscopy

6.2.1 Formulation

The electric field at position $\boldsymbol{r}$ in the fiber probe is expressed as

$$\boldsymbol{E}(\boldsymbol{r}) = \boldsymbol{E}_0(\boldsymbol{r}) + \sum_{i=1}^{N+M} \boldsymbol{T}^{\mathrm{t}}(\boldsymbol{r}, \boldsymbol{r}'_i)\alpha_i \boldsymbol{E}(\boldsymbol{r}'_i) , \tag{6.20}$$

by summing over all the dipole–dipole interactions among the $N+M$ spheres, i.e., by summing the multiple scattering effects between spheres. Here, $\boldsymbol{E}_0(\boldsymbol{r})$ is the electric field of the incident light and $\boldsymbol{T}^t(\boldsymbol{r}, \boldsymbol{r}')$ is the propagator given by

$$\boldsymbol{T}^{\mathrm{t}}(\boldsymbol{r}, \boldsymbol{r}'_i) = \boldsymbol{T}(\boldsymbol{r}, \boldsymbol{r}'_i) + \boldsymbol{T}_{\mathrm{M}}(\boldsymbol{r}, \boldsymbol{r}'_{i\mathrm{M}}) . \tag{6.21}$$

Replacing $\boldsymbol{E}(\boldsymbol{r}'_i)$ by $\boldsymbol{E}_0(\boldsymbol{r}'_i)$ on the right-hand side of (6.20), this equation is approximated as[4]

$$\boldsymbol{E}(\boldsymbol{r}) = \boldsymbol{E}_0(\boldsymbol{r}) + \sum_{i=1}^{N+M} \boldsymbol{T}^{\mathrm{t}}(\boldsymbol{r}, \boldsymbol{r}'_i)\alpha_i \boldsymbol{E}_0(\boldsymbol{r}'_i) . \tag{6.22}$$

Several quantities can be derived from this equation [6.3]. For example, the measurable intensity of the scattered light can be derived as

$$I(\theta) = \int_0^{\theta} R^2 \mathrm{d}\Omega |\boldsymbol{E}_{\mathrm{f}}(\boldsymbol{R} + \boldsymbol{R}_{\mathrm{P}})|^2 , \tag{6.23}$$

where θ is the cone angle of the fiber probe. This is the intensity of the light propagatedto position $\boldsymbol{R}$ in the fiber probe after the optical near field is converted to scattered light at the position $\boldsymbol{R}_{\mathrm{P}}$ of the probe tip.[5] It is thus

[4] This is called Born approximation, in which the generated electric field $\boldsymbol{E}(\boldsymbol{r}'_i)$ on the right-hand side is replaced by $\boldsymbol{E}_0(\boldsymbol{r}'_i)$ for the incident light. This approximation is valid if the magnitude of the second term on the right-hand side of (6.20) is assumed to be sufficiently smaller than the magnitude of the first term.

expressed as

$$\boldsymbol{E}_{\mathrm{f}}(\boldsymbol{R}+\boldsymbol{R}_{\mathrm{P}}) = \boldsymbol{T}_1(\boldsymbol{R}+\boldsymbol{R}_{\mathrm{P}}, \boldsymbol{R}_{\mathrm{P}})\alpha_{\mathrm{P}}\boldsymbol{E}(\boldsymbol{R}_{\mathrm{P}}) , \tag{6.24}$$

assuming that all the polarizabilities α_i have the same value α_{P}. In polar coordinates, $\mathrm{d}\Omega$ and $\boldsymbol{R}$ in (6.23) are given by

$$\mathrm{d}\Omega = \sin\theta'\mathrm{d}\theta'\mathrm{d}\phi \quad \text{and} \quad \boldsymbol{R} = (R\sin\theta'\cos\phi, R\sin\theta'\sin\phi, R\cos\theta') .$$

By using the explicit form of $\boldsymbol{T}_1(\boldsymbol{r},\boldsymbol{r}')$ in (6.14a), (6.24) is transformed to

$$\boldsymbol{E}_{\mathrm{f}}(\boldsymbol{R}+\boldsymbol{R}_{\mathrm{P}}) = -\frac{1}{4\pi\varepsilon_0}\frac{k^2}{R}\alpha_{\mathrm{P}}\boldsymbol{s} , \tag{6.25}$$

where we denote

$$\boldsymbol{s} = \begin{pmatrix} \sin^2\theta'\cos^2\phi - 1 & \sin^2\theta'\sin\phi\cos\phi & \sin\theta'\cos\theta'\cos\phi \\ \sin^2\theta'\sin\phi\cos\phi & \sin^2\theta'\sin^2\phi - 1 & \sin\theta'\cos\theta'\sin\phi \\ \sin\theta'\cos\theta'\cos\phi & \sin\theta'\cos\theta'\sin\phi & \cos^2\theta' - 1 \end{pmatrix} \boldsymbol{E}(\boldsymbol{R}_{\mathrm{P}}) . \tag{6.26}$$

Substituting (6.25) and (6.26) into (6.23), one obtains

$$\begin{aligned} I(\theta) &= \left(\frac{k^2}{4\pi\varepsilon_0}\right)^2 |\alpha_{\mathrm{P}}|^2 \int_0^{2\pi} \mathrm{d}\phi \int_0^{\theta} \sin\theta'\mathrm{d}\theta'|s|^2 \\ &= \left(\frac{k^2}{4\pi\varepsilon_0}\right)^2 |\alpha_{\mathrm{P}}|^2 \Big\{ \left[|E_x(\boldsymbol{R}_{\mathrm{P}})|^2 + |E_y(\boldsymbol{R}_{\mathrm{P}})|^2\right](16 - 15\cos\theta - \cos 3\theta) \\ &\quad + |E_z(\boldsymbol{R}_{\mathrm{P}})|^2(16 - 18\cos\theta + 2\cos 3\theta) \Big\} , \end{aligned} \tag{6.27}$$

where $E_x(\boldsymbol{R}_{\mathrm{P}})$, $E_y(\boldsymbol{R}_{\mathrm{P}})$, and $E_z(\boldsymbol{R}_{\mathrm{P}})$ are the components of $\boldsymbol{E}(\boldsymbol{R}_{\mathrm{P}})$ along the x-, y-, and z-axes, respectively.

For further discussion, the scattered light intensity I_E at the probe tip is defined as $I_E = |\boldsymbol{E}(\boldsymbol{R}_{\mathrm{P}})|^2$ with $\boldsymbol{E}(\boldsymbol{r})$ given by (6.22). Then, the relation between I_E and $I(\theta)$ of (6.27) is found to be

$$I_E = \frac{I(\theta = 90°)}{16(k^2/4\pi\varepsilon_0)^2|\alpha_{\mathrm{P}}|^2} . \tag{6.28}$$

The main advantages of the self-consistent method reviewed above are:

- It can be applied for an arbitrarily shaped three-dimensional probe and sample. This is because (6.22) is valid for an ensemble of spheres.
- The effect of polarization, i.e., the vectorial feature of the electric field, can be evaluated. This is because (6.11) introduced the electric field as a vector.

[5] All scattered light is assumed to be converted to the guided modes of the optical fiber.) The quantity $\boldsymbol{E}_{\mathrm{f}}(\boldsymbol{R}+\boldsymbol{R}_{\mathrm{P}})$ in this equation represents the electric field of the light propagated to the far-field position $\boldsymbol{R}$. [Note that this equation employs $\boldsymbol{T}_1(\boldsymbol{r},\boldsymbol{r}')$ of (6.14) because $k|\boldsymbol{R}+\boldsymbol{R}_{\mathrm{P}}| \gg 1$. It is the far-field component of (6.13), which has the largest value in the region $kR \gg 1$.]

- A range of experimental results can be discussed. This is because (6.22)–(6.28) treat the electric field in the fiber probe and the light intensity scattered by the fiber probe, which are measurable quantities.

However, the disadvantage is that numerical calculations with a long computation time are required to derive numerical results when Born approximation cannot be applied, and it is not easy to obtain physically intuitive concepts and perspectives.

6.2.2 Example Applications

In order to demonstrate the simplest application of the self-consistent method, the numbers N and M are fixed at unity, i.e., the sample and probe are assumed to be single spheres.

Dependence of I_E on the Polarization of the Incident Light

Figures 6.4a and b represent the values of I_E for s-polarized and p-polarized incident light, respectively. They are calculated from (6.28) in order to demonstrate the second advantage listed above. The sample and probe are assumed to be single spheres with radius 15 nm. These figures show that the values of I_E for s- and p-polarized incident light have minimum and maximum values at the center, respectively, consistent with the characteristics explained in Figs. 4.10 and 4.13. These features mean that the image is inverted in the case of s-polarized incident light, while it is not for p-polarized light.

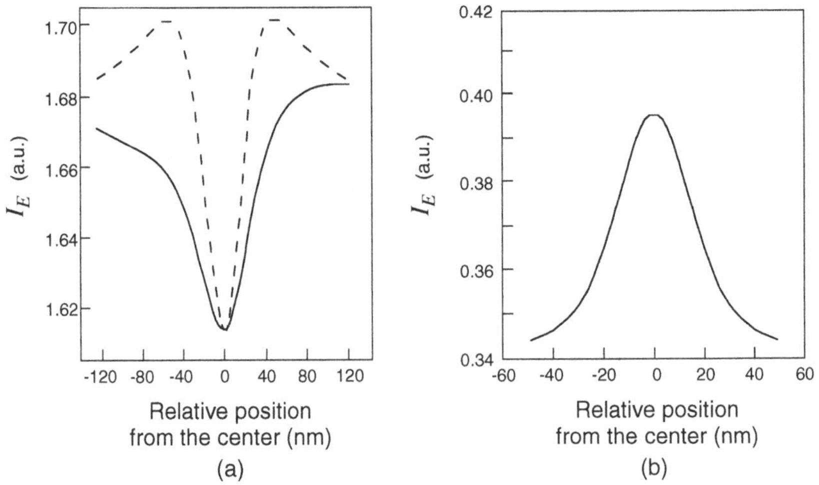

Fig. 6.4. Calculated values of I_E. The spherical sample and probe each have radius 15 nm. (**a**) The incident light is s-polarized. *Solid* and *broken curves* represent the results when the probe scans along the x- and y-axes, respectively. (**b**) The incident light is p-polarized

The solid and broken curves in Fig. 6.4a are the results when the probe scans along the x- and y-axes, respectively, as explained in Figs. 4.11a and b. The broken curve has two peaks at the edges of the sample, i.e., the edge effect depends on the direction of scanning, which explains the feature shown in Figs. 4.12b and c.

Dependence of $I(\theta)$ on the Cone Angle θ of the Probe

Figures 6.5a–d represent the values of $I(\theta)$ in the case of p-polarized incident light, calculated by using (6.23). Sample and probe sizes are the same as for Fig. 6.4. When $\theta \geq 45°$, the curves take the maximum value at the center,

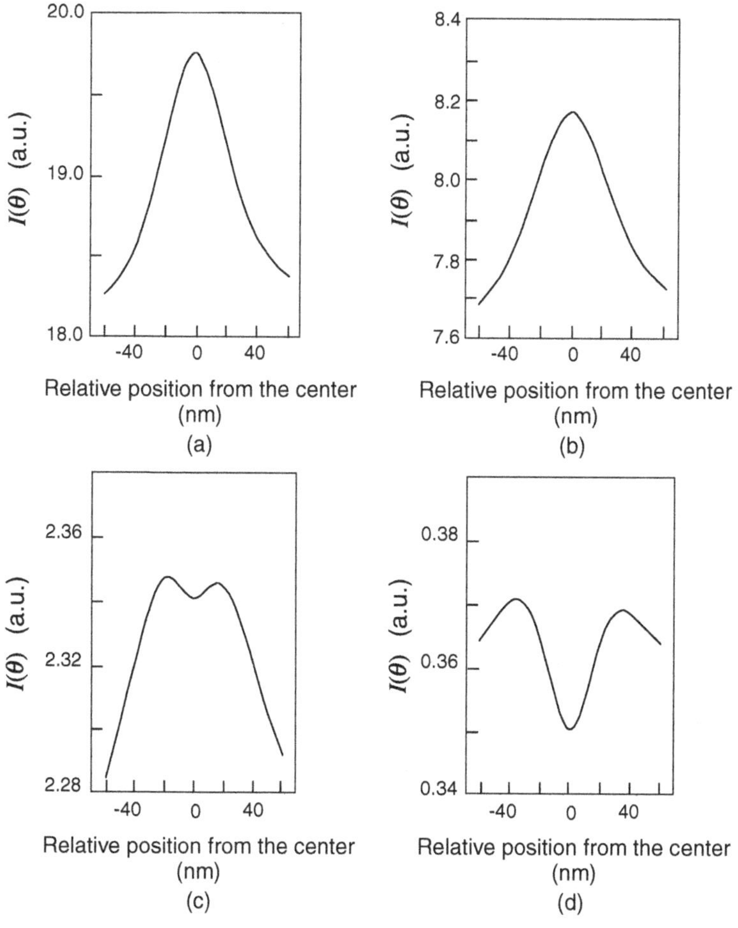

Fig. 6.5. Calculated values of $I(\theta)$ for p-polarized incident light. The spherical sample and probe each have radius 15 nm. The values of the cone angle θ are 60° (**a**), 45° (**b**), 30° (**c**), and 15° (**d**), respectively

corresponding to Fig. 6.4b. However, they have the minimum value at the center and peaks appear at the two edges of the sample for $\theta < 45°$. This corresponds to the polarization-dependent edge effect shown in Fig. 4.14. This effect can be explained as follows. Since the scattered light generated by the p-polarized light propagates along the x- and y-axes, detection efficiency is low when the light is collected by a probe with small θ fixed above the sample. Efficiency becomes higher when the probe is displaced from the top, whereupon peaks appear at the two edges. However, efficiency becomes less dependent on the position of the probe for larger θ, and the peaks disappear.

In the case of s-polarized incident light, features of the curves $I(\theta)$ are similar to those of I_E for s-polarized incident light, which have been shown in Fig. 6.4a. Although the value of $I(\theta)$ depends on θ, the profile of the curve $I(\theta)$ is almost independent of θ. This is because the electric force lines originating from the sample lie along the horizontal direction (i.e., the y-axis) above the sample, as shown in Fig. 4.10. The scattered light thus propagates along the z-axis.

Problems

Problem 6.1

Derive (6.12).

Problem 6.2

Derive (6.13).

Problem 6.3

Equation (A.28) of Appendix A represents the electric field $\boldsymbol{E}(\boldsymbol{r})$ generated by an electric dipole moment $\boldsymbol{p}$. Derive this equation using the propagator $\boldsymbol{T}(\boldsymbol{r}, \boldsymbol{r}')$ of (6.10).

Problem 6.4

Derive (6.15).

7 Picture of Optical Near Field Based on Electric Charges Induced on the Surface and Polarized Currents

Although the self-consistent method of Chap. 6 can deal with a sample and probe with arbitrary shapes, it is not straightforward to obtain physically intuitive concepts and perspectives because it relies on numerical analyses with long computation times to derive quantitative results. In order to overcome this difficulty, the present chapter transforms the basic formulas of electromagnetism and presents a novel theoretical model by introducing a dual vector potential and a scalar potential. This model is useful for a systematic analysis of the three cases listed in Sect. 4.1. In order to demonstrate this advantage, Sect. 7.1 describes the case in which the near-field condition is met, i.e., the sizes of the material systems under study and their separation are sufficiently smaller than the wavelength of the incident light. Section 7.2 describes the case of the quasi-near-field condition, i.e., the near-field condition is not met with sufficiently high accuracy.

7.1 Description under Near-Field Condition

This section describes Case 3 of Sect. 4.1, in which the near-field condition is met [7.1, 7.2].

7.1.1 Derivation of Electric Field Based on Static Electromagnetism

Consider detecting the electric field of light around sub-wavelength-sized dielectric materials in vacuum under the near-field condition. Comparing (A.15a) and (A.16a) of Appendix A, the polarization $\boldsymbol{P}(\boldsymbol{r},t)$ induced in the dielectric materials to be used as a sample for near-field optical microscopy can be expressed as

$$\boldsymbol{P}(\boldsymbol{r},t) = \left[\varepsilon(\boldsymbol{r}) - \varepsilon_0\right]\boldsymbol{E}(\boldsymbol{r},t) , \tag{7.1}$$

where the dielectric constant $\varepsilon(\boldsymbol{r})$ is

$$\varepsilon(\boldsymbol{r}) = \begin{cases} \varepsilon_1 & \text{inside the sample} , \\ \varepsilon_0 & \text{outside the sample} , \end{cases} \tag{7.2}$$

and ε_0 is the dielectric constant in vacuum. Since the time delay is negligible under the near-field condition ($kb < kr \ll 1$) as pointed out in Sect. 4.1, the light-scattering problem can be restricted to static electromagnetism. In particular, since true charges do not exist in the dielectric materials, the problem can be solved using Gauss's law, which is expressed by (A.3) of Appendix A. For the present discussion, the right-hand side of (A.3) is fixed at 0 (i.e., $\nabla\cdot\boldsymbol{E} = 0$), which means that the electric field is derived from Maxwell's equations neglecting the effect of time delay, i.e., neglecting the factor $e^{-i\omega t}$. Thus, it also means that the electric field is derived at $t = 0$.

Note that $\boldsymbol{D} = \varepsilon\boldsymbol{E}$ and substitute (A.16a) into $\nabla\cdot\boldsymbol{D} = 0$ in order to apply Gauss's law to the dielectric. Then, the equation $\nabla\cdot\boldsymbol{E}(\boldsymbol{r}) = (-1/\varepsilon_0)\nabla\cdot\boldsymbol{P}(\boldsymbol{r})$ is derived. Next, using (7.1) and the relation $\boldsymbol{E} = -\nabla\phi$ (where ϕ is a scalar potential), (A.3) is transformed to[1]

$$\begin{aligned}-\Delta\phi(\boldsymbol{r}) &= \frac{1}{\varepsilon_0}\nabla\cdot\boldsymbol{P}(\boldsymbol{r}) = \nabla\cdot\left[\frac{\varepsilon(\boldsymbol{r})-\varepsilon_0}{\varepsilon_0}\right]\nabla\phi(\boldsymbol{r}) \\ &= \nabla\left[\frac{\varepsilon(\boldsymbol{r})-\varepsilon_0}{\varepsilon_0}\right]\cdot\nabla\phi(\boldsymbol{r}) + \left[\frac{\varepsilon(\boldsymbol{r})-\varepsilon_0}{\varepsilon_0}\right]\Delta\phi(\boldsymbol{r}) . \qquad (7.3)\end{aligned}$$

Rearranging this equation, one obtains

$$-\Delta\phi(\boldsymbol{r}) = \frac{\nabla\varepsilon(\boldsymbol{r})}{\varepsilon(\boldsymbol{r})}\cdot\nabla\phi(\boldsymbol{r}) , \qquad (7.4)$$

which corresponds to Poisson's equation as in (A.4) of Appendix A. Comparing with the right-hand side of (A.4), it is found that the right-hand side of (7.4) corresponds to the electric charge density, i.e., the (surface density of electric charges induced on the surface)/ε_0, because the quantity $\nabla\varepsilon(\boldsymbol{r})$ is nonzero only on the dielectric surface. This nonzero value shows that the present electromagnetic system has a singularity on the surface, which corresponds to the boundary condition associated with Maxwell's equations. This means that the concept of the boundary is still valid under the near-field condition, while the time delay is neglected.

Noting that $\nabla\phi(\boldsymbol{r})$ is equal to $-\boldsymbol{E}(\boldsymbol{r})$ and employing Born approximation [i.e., $\boldsymbol{E}(\boldsymbol{r}) \approx \boldsymbol{E}_0(\boldsymbol{r})$, the electric field of the incident light], (7.4) can be approximated as

$$-\Delta\phi(\boldsymbol{r}) \approx \left(\frac{\varepsilon_1-\varepsilon_0}{\varepsilon_0}\right)\boldsymbol{n}_s\cdot\boldsymbol{E}_0(\boldsymbol{r})\int_S \delta^3(\boldsymbol{r}-\boldsymbol{s})\mathrm{d}^2 s , \qquad (7.5)$$

where $\boldsymbol{n}_s$ is the unit vector normal to the surface and directed outward, δ is the delta function, and $\boldsymbol{s}$ is the vector identifying the position on the dielectric surface.[2] It should be noted that the right-hand side of this equation includes

[1] See Problem 7.1

[2] Since the quantity $\{[\nabla\varepsilon(\boldsymbol{r})]/\varepsilon_0\}\cdot\nabla\phi(\boldsymbol{r})$ on the right-hand side of (7.4) is a product of distributions, it is transformed in order to reproduce the boundary condition associated with Maxwell's equations. The right-hand side of (7.5) represents the result of this transformation.

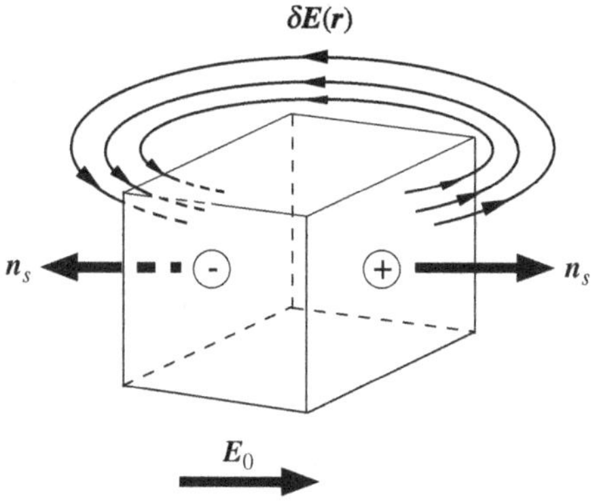

Fig. 7.1. Schematic representation of the electric field $\boldsymbol{E}_0$ of the incident light, the surface electric charges induced on a cubic piece of dielectric matter, and the electric lines of force illustrating the direction of the electric flux, where $\boldsymbol{n}_s$ is the unit vector normal to the surface and directed outward

the surface integral, where the region of integration is the whole dielectric surface.

Under this approximation, the electric field $\delta\boldsymbol{E}(\boldsymbol{r})(= -\nabla\phi)$ of the scattered light generated by electric charges induced on the surface can be derived from Gauss's law, i.e., its electric lines of force can be derived from (7.5). Figure 7.1 shows the surface electric charges induced on the dielectric cube surface and the electric lines of force representing the direction of the electric flux, which were illustrated by the method given above. This figure shows that the positive electric charges (identified by + in the figure) are induced on the surface with the unit vector $\boldsymbol{n}_s$ parallel to $\boldsymbol{E}_0$. The negative electric charges (identified by − in the figure) are on its rear surface. The electric lines of force are directed from positive to negative electric charges. The normalized light intensity is defined as

$$I(\boldsymbol{r}) \equiv \frac{|\boldsymbol{E}_0(\boldsymbol{r}) + \delta\boldsymbol{E}(\boldsymbol{r})|^2 - |\boldsymbol{E}_0(\boldsymbol{r})|^2}{|\boldsymbol{E}_0(\boldsymbol{r})|^2} = \frac{2\boldsymbol{E}_0(\boldsymbol{r})\cdot\delta\boldsymbol{E}(\boldsymbol{r}) + |\delta\boldsymbol{E}(\boldsymbol{r})|^2}{|\boldsymbol{E}_0(\boldsymbol{r})|^2} . \tag{7.6}$$

Because $|\boldsymbol{E}_0(\boldsymbol{r})| \gg |\delta\boldsymbol{E}(\boldsymbol{r})|$, the principal term in the numerator on the second line of this equation is $2\boldsymbol{E}_0(\boldsymbol{r})\cdot\delta\boldsymbol{E}(\boldsymbol{r})$, which represents interference between the incident and scattered light.[3] It should be noted that the incident and scattered light interferes under the near-field condition. The interfering term $2\boldsymbol{E}_0(\boldsymbol{r})\cdot\delta\boldsymbol{E}(\boldsymbol{r})$ has a negative value if the directions of the electric fields

[3] In Cases 1 and 2 of Sect. 4.1, on the other hand, the scattered light intensities measured are expressed as $I(\boldsymbol{r}) = |\delta E(\boldsymbol{r})|^2/|E_0(\boldsymbol{r})|^2$. This is because the incident light does not illuminate the photodetector placed in the far field.

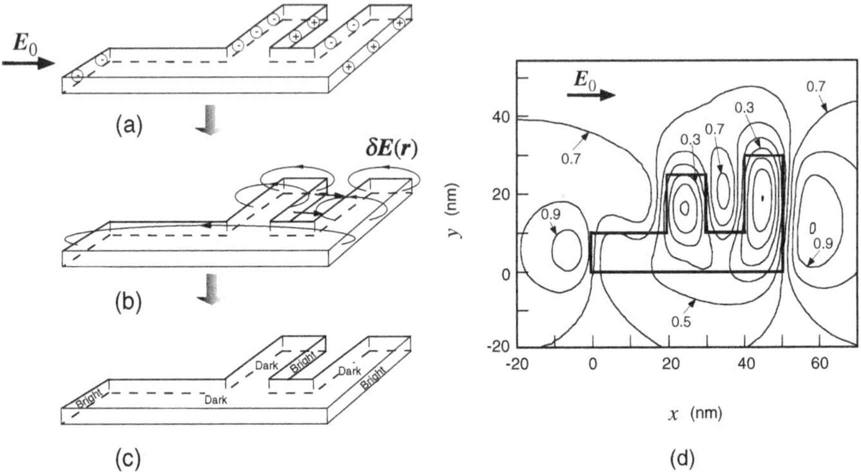

Fig. 7.2. Schematic explanation of how to derive and illustrate the spatial distribution of the light intensity $I(\boldsymbol{r})$ on the dielectric surface. (**a**), (**b**), and (**c**) correspond to steps (1), (2), and (3), respectively. (**d**) represents the result of numerical calculation based on (7.5) [7.3]. The value of I is given on each contour in order to compare the values in Fig. 1a of [7.4]. The relation between this value and $I(\boldsymbol{r})$ of (7.6) is $I(\boldsymbol{r}) = 0.25I - 0.17$

$\boldsymbol{E}_0(\boldsymbol{r})$ and $\delta\boldsymbol{E}(\boldsymbol{r})$ are opposite, and as a result, the intensity is lower than that of the background light.

As an example of the application of this formulation, the spatial distribution of the light intensity $I(\boldsymbol{r})$ on the dielectric surface is derived and illustrated by the following steps (refer to Fig. 7.2):

(1) The surface density of the surface electric charges is derived from the quantity $\left[(\varepsilon_1 - \varepsilon_0)/\varepsilon_0\right]\boldsymbol{n}_s\cdot\boldsymbol{E}_0(\boldsymbol{r})$ and illustrated (see Fig. 7.2a).
(2) The spatial distribution of the electric field $\delta\boldsymbol{E}(\boldsymbol{r})$ of the scattered light is derived and illustrated (see Fig. 7.2b). Note that the electric lines of force representing the electric field $\delta\boldsymbol{E}(\boldsymbol{r})$ are directed from the positive to negative surface electric charges.
(3) The spatial distribution of the intensity $I(\boldsymbol{r})$ is derived and illustrated (see Fig. 7.2c). The positive or negative sign of the scalar product $\boldsymbol{E}_0(\boldsymbol{r})\cdot\delta\boldsymbol{E}(\boldsymbol{r})$ determines whether the intensity $I(\boldsymbol{r})$ is higher or lower than that of the background light.

Equation (7.5) is used for numerical analysis to illustrate this spatial distribution quantitatively. For reference, Fig. 7.2d shows the result of such a quantitative illustration [7.3], which agrees with the result obtained by the self-consistent method presented in Chap. 6 [7.4]. An advantage of the present method is that the computation time is much shorter than by the self-consistent method.[4]

7.1.2 Signal Intensity Detected by a Fiber Probe

In this section we discuss collection-mode near-field optical microscopy.[5] Then (7.6) is replaced by

$$I_{\perp}(\boldsymbol{r}) \equiv \frac{|[\boldsymbol{E}_0(\boldsymbol{r}) + \delta\boldsymbol{E}(\boldsymbol{r})]_{\perp}|^2 - |\boldsymbol{E}_{0\perp}(\boldsymbol{r})|^2}{|\boldsymbol{E}_0(\boldsymbol{r})|^2}$$
$$= \frac{2\boldsymbol{E}_{0\perp}(\boldsymbol{r})\cdot\delta\boldsymbol{E}_{\perp}(\boldsymbol{r}) + |\delta\boldsymbol{E}_{\perp}(\boldsymbol{r})|^2}{|\boldsymbol{E}_0(\boldsymbol{r})|^2}, \tag{7.7}$$

where the symbol $\perp$ represents the direction normal to the fiber axis. We now discuss the dependence of $I_{\perp}(\boldsymbol{r})$ on the polarization of the light incident upon the dielectric sample.

S-Polarized Incident Light

The direction of the electric field $\boldsymbol{E}_0(\boldsymbol{r})$ of the *s*-polarized incident light is normal to the fiber axis, as shown in the upper part of Fig. 7.3a, i.e., $\boldsymbol{E}_{0\perp}(\boldsymbol{r}) = \boldsymbol{E}_0(\boldsymbol{r})$. Noting also that $|\boldsymbol{E}_{0\perp}(\boldsymbol{r})| \gg |\delta\boldsymbol{E}_{\perp}(\boldsymbol{r})|$, (7.7) can be approximated by $I_{\perp}(\boldsymbol{r}) \approx 2\boldsymbol{E}_0(\boldsymbol{r})\cdot\delta\boldsymbol{E}_{\perp}(\boldsymbol{r})/|\boldsymbol{E}_0(\boldsymbol{r})|^2$. The quantity $I_{\perp}(\boldsymbol{r})$ reaches its minimum value above the dielectric sample, as shown in the lower part of Fig. 7.3a, because $\delta\boldsymbol{E}_{\perp}(\boldsymbol{r})$ is antiparallel to $\boldsymbol{E}_0(\boldsymbol{r})$. As a result, the cross-sectional profile of $I_{\perp}(\boldsymbol{r})$ has a negative value with minimum at the center. This represents image inversion, which agrees with the characteristics described in Sect. 4.2.4.

[4] Agreement can be confirmed by comparing Fig. 7.2d with Fig. 1a of [7.4]. The numerical values I on the curves in Fig. 7.2d are consistent with those in Fig. 1a of [7.4]. The relation between I of Fig. 7.2d and $I(\boldsymbol{r})$ of (7.6) is $I(\boldsymbol{r}) = 0.25I - 0.17$. Thus the value of $I(\boldsymbol{r})$ is positive in the case $I \geq 0.68$, which means that the light intensity is higher than the background and the image looks brighter. Figures 1b–d of [7.4] show the results for a different case from Fig. 1a, which also agree with the results obtained by the present method [7.3]. The computation time required to illustrate the curves in Fig. 7.2d was only about 1 second on a personal computer (student version). In contrast to this, [7.4] explains that the computation time for Fig. 1a was as long as 5.5 min using the IBM RISC system/6000 model work station.

[5] We assume that the polarization of the light guided through the fiber is normal to the fiber axis, i.e., that the electric field of the guided light wave is transverse to the fiber axis, so that we have what is called a TE wave.

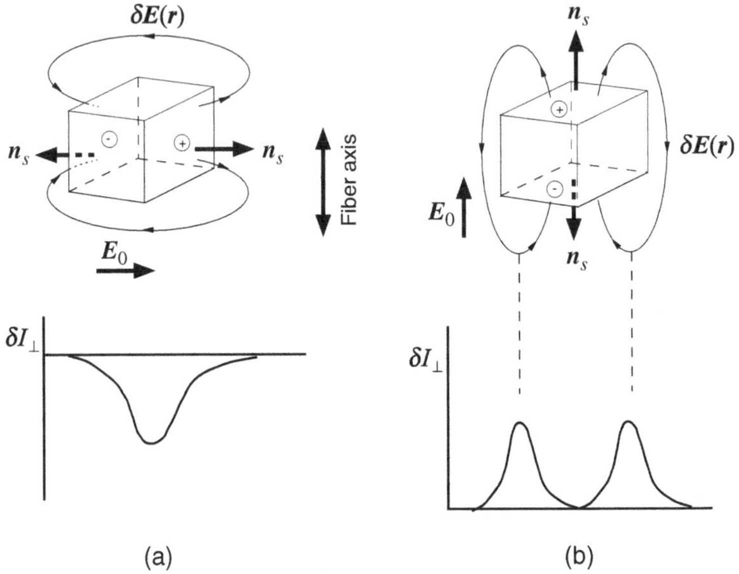

Fig. 7.3. Schematic explanation of the dependence of $I_{\perp}(\boldsymbol{r})$ on the polarization of the incident light. (**a**) and (**b**) show *s*- and *p*-polarized incident light, respectively

P-Polarized Incident Light

The direction of the electric field $\boldsymbol{E}_0(\boldsymbol{r})$ of the *p*-polarized incident light is parallel to the fiber axis, as shown in the upper part of Fig. 7.3b, i.e., $\boldsymbol{E}_{0\perp}(\boldsymbol{r}) = 0$ and thus (7.7) is $I_{\perp}(\boldsymbol{r}) = |\delta\boldsymbol{E}_{\perp}(\boldsymbol{r})|^2/|\boldsymbol{E}_0(\boldsymbol{r})|^2$. It is found from this figure that the value of $\delta\boldsymbol{E}_{\perp}(\boldsymbol{r})$ is also 0 on the top of the dielectric sample. On the other hand, $\delta\boldsymbol{E}(\boldsymbol{r})$ is horizontal at the edge of the dielectric sample, at which the value of $|\delta\boldsymbol{E}_{\perp}(\boldsymbol{r})|$ takes the maximum. As a result, the cross-sectional profile of $I_{\perp}(\boldsymbol{r})$ has peaks at the two edges, as shown in the lower part of Fig. 7.3b. It represents the polarization-dependent edge effect, which agrees with the characteristics demonstrated by Figs. 4.13 and 4.14.

7.2 Systematic Description of Optical Near and Far Fields

This section presents a physically intuitive concept for describing the quasi-near-field condition ($kb \leq kr \leq 1$), which is the transition from near-field ($kb < kr \ll 1$) to far-field ($1 \ll kb \ll kr$) conditions. In other words, the quasi-near-field condition describes the case of relatively poor conditions for near-field optics, in which the sizes of the sample and probe are not sufficiently small. The effect of the time delay cannot be neglected under this condition.

7.2.1 Dual Vector Potential

Conventional optics defines the vector potential $\boldsymbol{A}$ and the scalar potential ϕ by (A.7) and (A.18) of Appendix A, respectively. They have been widely used to solve electromagnetic problems for matter with a magnetic response ($\boldsymbol{M} \neq 0$) and without electric response ($\boldsymbol{P} = 0$).[6] For such matter, (A.19) is valid with $\boldsymbol{A}(\boldsymbol{r}, t) = 0$ and $\phi(\boldsymbol{r}, t) = 0$. Further, the magnetic current $\nabla \times \boldsymbol{M}(\boldsymbol{r}, t)$ on the right-hand side of (A.19) acts as the source for the vector potential $\boldsymbol{A}(\boldsymbol{r}, t)$.

The problem to be solved in this section concerns matter which is opposite to that described above, i.e., matter with an electric response ($\boldsymbol{P} \neq 0$) and without magnetic response ($\boldsymbol{M} = 0$). Noting the duality of Maxwell's equations [(A.12)–(A.16) of Appendix A], it is convenient to introduce a vector potential $\boldsymbol{C}$ and a scalar potential χ for the system composed of dielectrics and light. They are defined by

$$\boldsymbol{D} = \nabla \times \boldsymbol{C} , \tag{7.8a}$$

$$\boldsymbol{H} = \frac{\partial \boldsymbol{C}}{\partial t} + \nabla \chi . \tag{7.8b}$$

Replacing $\boldsymbol{A}$, ϕ, and $\boldsymbol{M}$ in (A.19) by $\boldsymbol{C}$, χ, and $\boldsymbol{P}$, respectively, one obtains a wave function

$$\nabla \times \nabla \times \boldsymbol{C}(\boldsymbol{r}, t) + \varepsilon_0 \mu_0 \frac{\partial^2 \boldsymbol{C}(\boldsymbol{r}, t)}{\partial t^2} = \nabla \times \boldsymbol{P}(\boldsymbol{r}, t) , \tag{7.9}$$

where

$$\nabla \cdot \boldsymbol{C}(\boldsymbol{r}, t) = 0 , \tag{7.10a}$$

$$\chi(\boldsymbol{r}, t) = 0 . \tag{7.10b}$$

Equation (7.9) shows that the source of the vector potential $\boldsymbol{C}(\boldsymbol{r}, t)$ is the quantity $\nabla \times \boldsymbol{P}(\boldsymbol{r}, t)$. This is called the polarized current and corresponds to the magnetic current $\nabla \times \boldsymbol{M}(\boldsymbol{r}, t)$ of the right-hand side of (A.19). Thus, after substituting the polarization $\boldsymbol{P}(\boldsymbol{r}, t)$ of (7.1) into (7.9), (7.8) is substituted into the relation $\boldsymbol{P} = \{[\varepsilon(\boldsymbol{r}) - \varepsilon_0]/\varepsilon(\boldsymbol{r})\}\boldsymbol{D}$, which is derived from (7.1) and (A.15a). As a result, the right-hand side of (7.9) is transformed to[7]

$$\begin{aligned}
\nabla \times \boldsymbol{P} &= \nabla \times \left\{ \left[1 - \frac{\varepsilon_0}{\varepsilon(\boldsymbol{r})} \right] \nabla \times \boldsymbol{C} \right\} \\
&= \left[1 - \frac{\varepsilon_0}{\varepsilon(\boldsymbol{r})} \right] \nabla \times \nabla \times \boldsymbol{C} + \nabla \left[1 - \frac{\varepsilon_0}{\varepsilon(\boldsymbol{r})} \right] \times \nabla \times \boldsymbol{C} \\
&= \left[1 - \frac{\varepsilon_0}{\varepsilon(\boldsymbol{r})} \right] \nabla \times \nabla \times \boldsymbol{C} + \left[\frac{\varepsilon_0 \nabla \varepsilon(\boldsymbol{r})}{\varepsilon^2(\boldsymbol{r})} \right] \times \nabla \times \boldsymbol{C} .
\end{aligned} \tag{7.11}$$

---

[6] It is not easy to find this kind of matter, except for superconductors.

[7] See Problem 7.2

Further, using $\partial/\partial t = -\mathrm{i}\omega$ and taking the value at $t = 0$ as in the previous section, (7.9) is transformed to

$$\begin{aligned} &\nabla \times \nabla \times \boldsymbol{C}(\boldsymbol{r}) - k^2\boldsymbol{C}(\boldsymbol{r}) \\ &\quad = \left[1 - \frac{\varepsilon_0}{\varepsilon(\boldsymbol{r})}\right] \nabla \times \nabla \times \boldsymbol{C}(\boldsymbol{r}) + \left[\frac{\varepsilon_0 \nabla\varepsilon(\boldsymbol{r})}{\varepsilon^2(\boldsymbol{r})}\right] \times \nabla \times \boldsymbol{C}(\boldsymbol{r}) . \end{aligned} \tag{7.12}$$

After rearranging several terms, one obtains

$$\nabla \times \nabla \times \boldsymbol{C}(\boldsymbol{r}) - k^2\boldsymbol{C}(\boldsymbol{r}) = -\boldsymbol{V}_{\mathrm{s}}(\boldsymbol{r}) - \boldsymbol{V}_{\mathrm{v}}(\boldsymbol{r}) , \tag{7.13}$$

where

$$\boldsymbol{V}_{\mathrm{s}}(\boldsymbol{r}) = -\frac{\nabla\varepsilon(\boldsymbol{r})}{\varepsilon(\boldsymbol{r})} \times \nabla \times \boldsymbol{C}(\boldsymbol{r}) , \tag{7.14a}$$

$$\boldsymbol{V}_{\mathrm{v}}(\boldsymbol{r}) = -\left[\frac{\varepsilon(\boldsymbol{r})}{\varepsilon_0} - 1\right] k^2\boldsymbol{C}(\boldsymbol{r}) . \tag{7.14b}$$

Equation (7.13) shows that the sources of the electromagnetic field are $\boldsymbol{V}_{\mathrm{s}}(\boldsymbol{r})$ and $\boldsymbol{V}_{\mathrm{v}}(\boldsymbol{r})$, whose physical identities are:

$\boldsymbol{V}_{\mathrm{s}}(\boldsymbol{r})$: surface polarized current, whose value is determined in such a way that the boundary condition associated with Maxwell's equation is satisfied.

$\boldsymbol{V}_{\mathrm{v}}(\boldsymbol{r})$: volume polarized current, which represents the effect of time delay because (7.14) depends on the wave number k.[8]

Equations (7.9)–(7.14) are valid for conditions ranging from the near field ($kb < kr \ll 1$) to the far field ($1 \ll kb \ll kr$), and can thus be applied to general systems composed of dielectrics and light.

7.2.2 Dual Ampere Law

In this section we investigate near-field ($kb < kr \ll 1$) and quasi-near-field ($kb \le kr \le 1$) conditions, which correspond to the cases of $kb \ll 1$ and $kb \le 1$, respectively, for the size of the dielectric matter.

Near-Field Condition

By applying Born approximation, i.e., by substituting $\boldsymbol{C}_0$ for the incident light into $k^2\boldsymbol{C}(\boldsymbol{r})$, $\boldsymbol{V}_{\mathrm{s}}(\boldsymbol{r})$, and $\boldsymbol{V}_{\mathrm{v}}(\boldsymbol{r})$, equation (7.13) reduces to

$$\nabla \times \boldsymbol{D}(\boldsymbol{r}) = -\boldsymbol{V}_{\mathrm{s}0}(\boldsymbol{r}) - \boldsymbol{V}_{\mathrm{v}0}(\boldsymbol{r}) + k^2\boldsymbol{C}_0(\boldsymbol{r}) , \tag{7.15}$$

[8] The effect of time delay is also contained in the second term $-k^2\boldsymbol{C}(\boldsymbol{r})$ on the left-hand side of (7.13). It represents the effect of diffraction.

where (7.8) was used to derive the left-hand side. The suffix 0 on $V_{s0}(\boldsymbol{r})$ and $\boldsymbol{V}_{v0}(\boldsymbol{r})$ indicates that $\boldsymbol{C}$ is replaced by $\boldsymbol{C}_0$ in (7.14a) and (7.14b), respectively. Because $kb \ll 1$, the quantities $\boldsymbol{V}_{v0}(\boldsymbol{r})$ and $k^2\boldsymbol{C}_0(\boldsymbol{r})$ can be neglected and removed from (7.15). Representing $\boldsymbol{D}(\boldsymbol{r})$ as $\boldsymbol{D}_0(\boldsymbol{r}) + \delta\boldsymbol{D}(\boldsymbol{r})$ and substituting the relation $\nabla \times \boldsymbol{D}_0(r) = k^2\boldsymbol{C}_0(\boldsymbol{r})$ into (7.15), one obtains

$$\nabla \times \delta\boldsymbol{D}(\boldsymbol{r}) = -\boldsymbol{V}_{s0}(\boldsymbol{r}) \,. \tag{7.16}$$

This equation means that the surface polarized current is the source of the electric flux density, and hence represents the dual Ampere law.[9]

Since the right-hand side of (7.16) contains a negative sign, the directions of $\boldsymbol{V}_{s0}(\boldsymbol{r})$ and $\delta\boldsymbol{D}(\boldsymbol{r})$ correspond to those of the thumb and four fingers of the left hand, respectively. Noting that $\varepsilon_0 \times$ (surface polarized current) $\approx$ $(\varepsilon_1/\varepsilon_0 - 1)\boldsymbol{n}_s \times \boldsymbol{D}_0$, $\boldsymbol{V}_{s0}(\boldsymbol{r})$ can be approximated as

$$\begin{aligned} \boldsymbol{V}_{s0}(\boldsymbol{r}) &\approx -\left[\frac{\nabla\varepsilon(\boldsymbol{r})}{\varepsilon(\boldsymbol{r})}\right] \times \nabla \times \boldsymbol{C}_0(\boldsymbol{r}) \\ &\approx \left(\frac{\varepsilon_1}{\varepsilon_0} - 1\right) \boldsymbol{n}_s \times \boldsymbol{D}_0 \int_S \delta^3(\boldsymbol{r} - \boldsymbol{s})\mathrm{d}^2 s \,, \end{aligned} \tag{7.17}$$

where a transformation similar to the one from (7.4) to (7.5) was used. As an example, Fig. 7.4a shows the relation between $\boldsymbol{n}_s$, $\boldsymbol{D}_0$, and the surface polarized current for the dielectric cube at $t = 0$. The surface polarized current is represented by bold arrows normal to $\boldsymbol{n}_s$ and $\boldsymbol{D}_0$ and flows on the surface of the cube. As a result of this flow, $\delta\boldsymbol{D}(r)$ is generated along the direction given by (7.16), which is shown by solid curves in this figure. That is, the thumb of the left hand represents the direction of $\delta\boldsymbol{D}(\boldsymbol{r})$, while the other four fingers represent the polarized current as shown in Fig. 7.4b. The direction of $\delta\boldsymbol{D}(\boldsymbol{r})$ is the same as the direction of $\delta\boldsymbol{E}(\boldsymbol{r})$ in Fig. 7.1, which gives the same spatial distribution for the normalized light intensity $I(\boldsymbol{r})$ as was obtained by the method of Sect. 7.1.

As an example of numerical analysis, Fig. 7.4c shows the calculated results for the dielectric cube with size $kb = 0.01$ and dielectric constant $\varepsilon_1/\varepsilon_0 = 2.25$. This figure illustrates the spatial distribution of $I(\boldsymbol{r})$ at height $0.2b$ above the dielectric surface [7.2], which agrees with the result calculated by the method of Sect. 7.1.

The profile of the spatial distribution of $I(\boldsymbol{r})$ is independent of kb, as long as the condition $kb \ll 1$ is met and time delay effects can be neglected. This means that the phenomenon observed under near-field conditions is

[9] Equation (A.13) for the conventional Ampere law represents the magnetic field generated by an electric current. The generated magnetic field is directed along the axis of rotation of the clockwise screw if the electric current is along the driving direction of the screw. Thus the directions of the electric current and magnetic field correspond to those of the thumb and four fingers of the right hand, respectively.

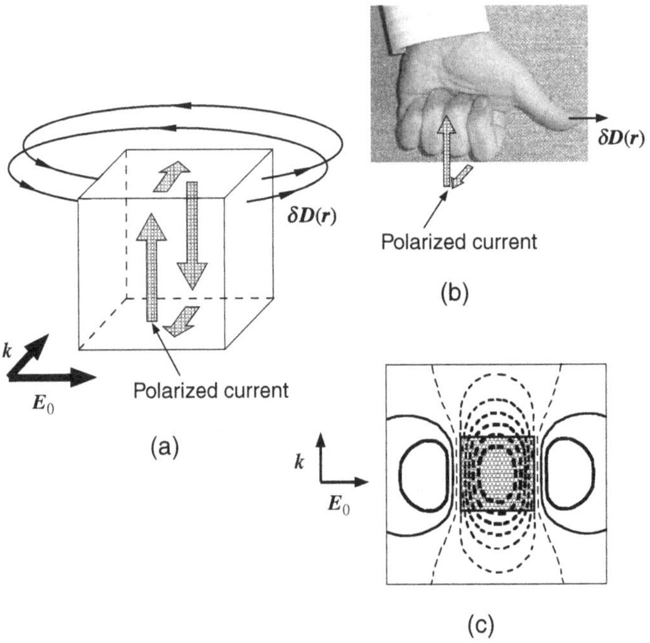

Fig. 7.4. Polarized current, electrical flux density, and spatial distribution of $I(\boldsymbol{r})$ on the surface of a dielectric cube. (**a**) *Bold arrows* and *solid curves* represent the polarized current and the electric flux density $\delta\boldsymbol{D}(\boldsymbol{r})$, respectively. (**b**) Photograph of a left hand. The thumb and other four fingers represent the directions of $\delta\boldsymbol{D}(\boldsymbol{r})$ and polarized current, respectively. (**c**) Calculated results of the spatial distribution of $I(\boldsymbol{r})$ at height $0.2b$ from the surface of the dielectric cube, where $kb = 0.01$ and $\varepsilon_1/\varepsilon_0 = 2.25$. The *square* at the center represents the position of the cube. *Solid* and *broken curves* represent $I(\boldsymbol{r}) > 0$ and $I(\boldsymbol{r}) < 0$, respectively. The *thicker curve* represents the larger absolute value of $I(\boldsymbol{r})$

independent of the wavelength of the incident light and thus free from the diffraction limit.

Quasi-Near-Field Condition

Figure 7.5a shows the spatial distribution of $I(\boldsymbol{r})$ calculated without neglecting the quantities $\boldsymbol{V}_{\mathrm{v}0}(\boldsymbol{r})$ and $k^2\boldsymbol{C}_0(\boldsymbol{r})$ in (7.15) [7.2], where $kb = 1.00$. Other numerical values used are the same as those for Fig. 7.4c. This figure shows that $I(\boldsymbol{r}) < 0$ in front of the cube, i.e., the light intensity is lower than the incident light intensity. This means that a kind of shadow appears in front of the cube due to the effect of time delay.

In Fig. 7.5b, the dielectric cube is replaced by a slab (thickness b, dielectric constant ε_1) placed on the xy-plane in order to obtain a more intuitive understanding [7.2]. The direction of $\boldsymbol{k}$ (i.e., the direction of the incident light

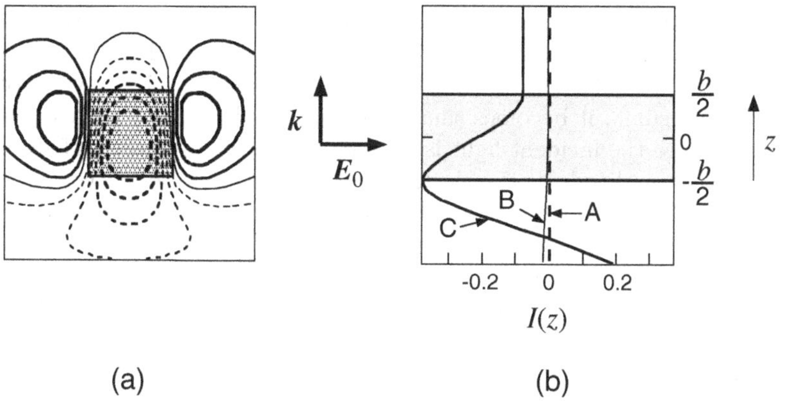

Fig. 7.5. Spatial distribution of $I(\boldsymbol{r})$ on the surface of the dielectric cube and its schematic explanation. (**a**) Calculated results for $I(\boldsymbol{r})$, where $kb = 1.00$. Other numerical values are the same as those of Fig. 7.4c. The *square* at the center represents the position of the cube. *Solid* and *broken curves* represent the contours with $I(\boldsymbol{r}) > 0$ and $I(\boldsymbol{r}) < 0$, respectively. The *thicker curve* represents the larger absolute value of $I(\boldsymbol{r})$. (**b**) Calculated results for the z-axis dependence of the intensity $I(z)$ for a dielectric slab (thickness b, dielectric constant ε_1). The slab is placed on the xy-plane and the wave vector $\boldsymbol{k}$ of the incident light is parallel to the z-axis. The surroundings are assumed to be a vacuum with dielectric constant ε_0 ($< \varepsilon_1$). The direction of the electric field $\boldsymbol{E}_0$ of the incident light is assumed to be parallel to the xy-plane. The *shaded band* represents the position of the slab (i.e., $-b/2 \leq z \leq b/2$), where $b = 1.0$ and $\varepsilon_1 = 2.25\varepsilon_0$. Curves A, B, and C are the results for $kb = 0.01$, 0.10, and 1.00, respectively

propagation) is parallel to the z-axis. The surroundings are assumed to be a vacuum with dielectric constant ε_0 ($< \varepsilon_1$). The direction of the electric field $\boldsymbol{E}_0(\boldsymbol{r})$ of the incident light is assumed to be parallel to the xy-plane. Noting that the value of the surface polarized current $\boldsymbol{V}_{\mathrm{s}}(\boldsymbol{r})$ of (7.14) is 0 in this case, while those of the volume polarized current $\boldsymbol{V}_{\mathrm{v}}(\boldsymbol{r})$ of (7.14) and the effect of time delay $k^2\boldsymbol{C}(\boldsymbol{r})$ are nonzero, equation (7.14) transforms to

$$\frac{\partial^2 s(z)}{\partial z^2} = -\frac{\varepsilon(z)}{\varepsilon_0} k^2 s(z) \,, \tag{7.18}$$

where the quantity $s(z)$ is defined by $\boldsymbol{C}(\boldsymbol{r}) = \nabla \times \boldsymbol{y}_{\mathrm{s}}(z)$, $\boldsymbol{y}$ is a unit vector along the y-axis, and

$$\varepsilon(z) = \begin{cases} \varepsilon_1 & -b/2 \leq z \leq b/2 \quad \text{(inside the slab)} \,, \\ \varepsilon_0 & z < -b/2, b/2 < z \quad \text{(outside the slab)} \,. \end{cases} \tag{7.19}$$

It is straightforward to solve this equation by noting the continuity at the slab surface ($z = \pm b/2$). Since $s(z)$ is proportional to the electric field amplitude, the normalized light intensity $I(z)$ can be derived if the electric field is replaced by $s(z)$.

Figure 7.5b shows the calculation results for $b = 1.0$ and $\varepsilon_1/\varepsilon_0 = 2.25$. Curves A, B, and C represent the value of $I(z)$ for $kb = 0.01$, 0.10, and 1.00, respectively. Although the value of $I(z)$ is independent of z if the value of kb is sufficiently small, it becomes smaller in front of the slab by increasing kb. This is because the incident light is reflected on the front surface of the slab and thus the amplitude of the electric field is decreased around the surface. This is analogous to a conventional optical phenomenon, i.e., a decrease in light intensity due to destructive interference between the incident light and the phase-inverted reflected light in front of the material surface.

Although it is not easy to evaluate the effect of the time delay in the case of a dielectric cube, the spatial distribution $I(r)$ of Fig. 7.5a also contains the effect of reflection, as demonstrated in Fig. 7.5b. This means that the light intensity is lower in front of the cube under the quasi-near-field condition, because the surface effect of (7.16) and the effect of time delay exist due to reflection. As a result, the shadow appears.

To summarize the above discussions, the characteristics of the spatial distribution of light intensity can be understood as a superposition of the boundary effect governed by Ampere's law and the effect of time delay due to reflection. By increasing the material size from the near field to the quasi-near-field condition, the front surface of the matter becomes darker. By further increasing the material size, the quasi-near-field condition can no longer be mct. In this case, the shadow appears behind the matter, which is a conventional optical phenomenon.

Problems

Problem 7.1

Transform the first line of (7.3) and derive the second line.

Problem 7.2

Transform the first line of (7.11) and derive the second line.

8 Picture of Optical Near Field as a Virtual Cloud Around a Nanometric System Surrounded by a Macroscopic System

Previous chapters used classical electromagnetism to describe a nanometric system composed of a sample, a probe, and an optical near field. This chapter presents a quantum mechanical model based on a projection operator method to describe the interaction between nanometric material systems via an optical near field surrounded by a macroscopic system. This model can also be used to describe the interaction between an atom and a probe, and its application to atom photonics is discussed in Chap. 9. Appendices C and D provide supplementary explanations of the concepts to be used in this chapter. An outstanding advantage of this model is its ability to systematically describe the light–matter interactions in nanometric material and atomic systems. This is because the model is based on concepts developed in the fields of elementary particle physics, statistical mechanics, quantum chemistry, and quantum optics. Furthermore, the model provides an intuitive physical picture in which the localized optical near field can be described in the same way as an electron cloud localized around an atomic nucleus.

8.1 Basic Concept

In the case of a collection-mode near-field optical microscope, there is a substrate under a sample and incident light from the light source, as shown in Fig. 8.1a. Further, behind the probe tip, there is a tapered part, the main body of the fiber, and a photodetector. Figure 8.1b shows that the illumination mode has incident light, the main body of the fiber, and a tapered part behind the probe tip. Behind the sample, there is a substrate and a photodetector. In both modes, there is a macroscopic system (composed of incident light, substrate, the main body of the fiber, a tapered part, and a photodetector) around the nanometric system (composed of sample, probe tip, and optical near field). Therefore, it should be noted that the nanometric tip of the probe (hereafter referred to as the probe for simplicity) and the sample interact electromagnetically via an optical near field surrounded by the macroscopic system.

The nanometric system is called subsystem (N), while the surrounding macroscopic system is called subsystem (M) (the first letters of 'nanometric'

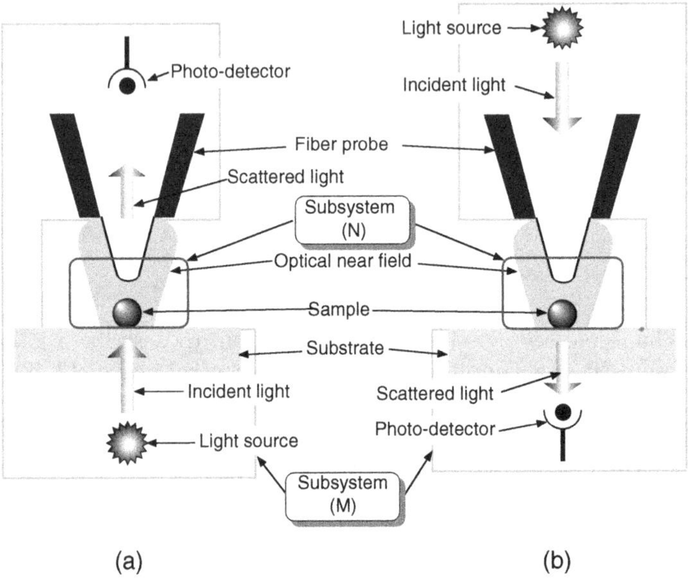

Fig. 8.1. Schematic configuration of subsystems (N) and (M) in a near-field optical microscope. (**a**) Collection mode. (**b**) Illumination mode

and 'macroscopic', respectively). All the effects and influences from subsystem (M) to subsystem (N) are described by modifying the magnitude of the electromagnetic interaction between the sample and probe. The idea is to avoid the complexity of describing all the behavior of subsystems (N) and (M) rigorously. Since we are interested only in the behavior of subsystem (N), it does not make sense to describe all the behavior.

In order to describe the quantum mechanical state of matter in subsystems (N) and (M), energy eigenstates of the sample and probe in the subsystem (N) are expressed as $|s\rangle$ and $|p\rangle$, respectively. Further, suffices g and e are added to express the ground and excited states, respectively, i.e., $|s_\mathrm{g}\rangle$, $|s_\mathrm{e}\rangle$, $|p_\mathrm{g}\rangle$, and $|p_\mathrm{e}\rangle$.

The tapered part, the main body of the fiber, and the substrate in the subsystem (M) are made of crystals, amorphous materials, and so on, in which there is incident and scattered light. Therefore, it is most reasonable to express the subsystem (M) as an exciton–polariton, which is a mixed state of material excitation and electromagnetic fields (see Appendix C for more detailed explanation). It can be represented by the state $|m_{(\mathrm{M})}; \boldsymbol{k}, \Omega(k)\rangle$, where $\boldsymbol{k}$ and $\Omega(k)$ are the wave vector and angular frequency of the exciton–polariton, respectively. The integer $m_{(\mathrm{M})}$ represents the number of quanta of the exciton–polariton. [The wave vector $\boldsymbol{k}$ and angular frequency $\Omega(k)$ of the exciton–polariton are proportional to the eigenvalues of its momentum $\hbar k$ and energy $\hbar\Omega(k)$, respectively.]

Since the sample or the probe is excited by electromagnetic interaction, the state of the subsystem (N) can be expressed as $|s_e\rangle|p_g\rangle$ or $|s_g\rangle|p_e\rangle$. By combining the states of the subsystem (N) and the vacuum state $|0_{(M)};\boldsymbol{k},\Omega(k)\rangle$ of the subsystem (M), we define a P-space $= \{|\phi_1\rangle, |\phi_2\rangle\}$, where $|\phi_1\rangle = |s_e\rangle|p_g\rangle|0_{(M)};\boldsymbol{k},\Omega(k)\rangle$ and $|\phi_2\rangle = |s_g\rangle|p_e\rangle|0_{(M)};\boldsymbol{k},\Omega(k)\rangle$. The supplementary space is called a Q-space, which is the space composed of other states. From now onwards, an arbitrary state $|\Psi\rangle$ of the total system (i.e., the P-space plus the Q-space) will be described in the P-space [8.1]. This method of description is called a projection operator method. (See Sect. D.4.1 for the definition and mathematical treatment of the projection operator.)

The reason why $|\phi_1\rangle$ and $|\phi_2\rangle$ contain the vacuum state $|0_{(M)};\boldsymbol{k},\Omega(k)\rangle$ is to introduce the effect of the subsystem (M) by eliminating its degree of freedom. As a result, subsystem (N) can be expressed as if it were independent of subsystem (M). This treatment is useful for deriving a consistent expression, in which subsystem (N) is regarded as being isolated from subsystem (M) while the functional form and the magnitude of effective interactions between the elements in subsystem (N) are affected by subsystem (M). (See also Sect. D.4.2 for the meaning of 'effective'.) By this treatment, one avoids having to consider all interactions between elements of subsystems (N) and (M).

8.2 Effective Interaction Between Sample and Probe

The quantum mechanical Hamiltonian for the interaction between a sample or a probe and electromagnetic fields is expressed as (see Sect. D.1 of Appendix D for derivation)

$$\hat{V} = -\frac{1}{\varepsilon_0}\left[\hat{\boldsymbol{p}}_S\cdot\hat{\boldsymbol{D}}(\boldsymbol{r}_S) + \hat{\boldsymbol{p}}_P\cdot\hat{\boldsymbol{D}}(\boldsymbol{r}_P)\right], \tag{8.1}$$

where ε_0 is the dielectric constant of vacuum. The subscript S in the first term stands for the sample, and $\hat{\boldsymbol{p}}_S$ is the quantum-mechanical operator representing an electric dipole induced in the sample, $\hat{\boldsymbol{r}}_S$ a vector representing the position of the sample, and $\hat{\boldsymbol{D}}(\boldsymbol{r}_S)$ the quantum mechanical operator for the electric flux density, respectively. The subscript P in the second term stands for the probe.

The operator $\hat{\boldsymbol{D}}(\boldsymbol{r}_S)$ is represented in terms of the annihilation and creation operators of the incident light, $\hat{a}_\lambda(\boldsymbol{k})$ and $\hat{a}_\lambda^\dagger(\boldsymbol{k})$, respectively. Then we have (see Sect. D.2 for details of the derivation)

$$\hat{\boldsymbol{D}}(\boldsymbol{r}) = \sum_{\boldsymbol{k}}\sum_{\lambda=1}^{2} \mathrm{i}\left(\frac{\varepsilon_0\hbar\omega_k}{2V}\right)^{1/2} \boldsymbol{e}_\lambda(\boldsymbol{k})\left[\hat{a}_\lambda(\boldsymbol{k})\mathrm{e}^{\mathrm{i}\boldsymbol{k}\cdot\boldsymbol{r}} - \hat{a}_\lambda^\dagger(\boldsymbol{k})\mathrm{e}^{-\mathrm{i}\boldsymbol{k}\cdot\boldsymbol{r}}\right], \tag{8.2}$$

where the two polarization states of the incident light are identified by the subscript $\lambda = 1, 2$. In these relations, $\boldsymbol{k}$ is the wave vector of the photon,

ω_k the angular frequency of the photon, V the volume of the space in which the electromagnetic fields exist, and $\boldsymbol{e}_\lambda(\boldsymbol{k})$ a unit vector representing the direction of polarization of the photon, respectively.

The incident light is affected by the structures and shapes of the materials in subsystem (M) because it transmits through subsystem (M) before it reaches the sample and probe of subsystem (N). Thus, the state of the incident light should be expressed as an exciton–polariton, which means that the operators $\hat{a}_\lambda(\boldsymbol{k})$ and $\hat{a}_\lambda^\dagger(\boldsymbol{k})$ in (8.2) should be replaced by the annihilation and creation operators, $\hat{\xi}_\lambda(\boldsymbol{k})$ and $\hat{\xi}_\lambda^\dagger(\boldsymbol{k})$, of the exciton–polariton. After this replacement, substitution of (8.2) into (8.1) gives (see Sect. D.3 for details of the derivation)

$$\hat{V} = -\mathrm{i}\left(\frac{\hbar}{2\varepsilon_0 V}\right)^{1/2} \sum_{\alpha=\mathrm{S}}^{\mathrm{P}} \left[\hat{B}(\boldsymbol{r}_\alpha) + \hat{B}^\dagger(\boldsymbol{r}_\alpha)\right] \sum_{k} \left[K_\alpha(\boldsymbol{k})\hat{\xi}(\boldsymbol{k}) - K_\alpha^*(\boldsymbol{k})\hat{\xi}^\dagger(\boldsymbol{k})\right], \tag{8.3}$$

where $\hat{B}(\boldsymbol{r}_\alpha)$ and $\hat{B}^\dagger(\boldsymbol{r}_\alpha)$ denote the annihilation and creation operators for the electronic excitation in the sample or probe ($\alpha = \mathrm{S}, \mathrm{P}$), and $K_\alpha(\boldsymbol{k})$ designates the coefficient representing the coupling strength between subsystem (N) and (M) while $K_\alpha^*(\boldsymbol{k})$ represents the complex conjugate of $K_\alpha(\boldsymbol{k})$. The coefficient $K_\alpha(\boldsymbol{k})$ is given by

$$\begin{aligned} K_\alpha(\boldsymbol{k}) &= \sum_{\lambda=1}^{2} \{\boldsymbol{p}_\alpha \cdot \boldsymbol{e}_\lambda(\boldsymbol{k})\} f(k) \mathrm{e}^{\mathrm{i}\boldsymbol{k}\cdot\boldsymbol{r}_\alpha} \\ &= \sum_{j=1}^{3} \sum_{\lambda=1}^{2} p_{\alpha j} \{\boldsymbol{e}_j \cdot \boldsymbol{e}_\lambda(\boldsymbol{k})\} f(k) \mathrm{e}^{\mathrm{i}\boldsymbol{k}\cdot\boldsymbol{r}_\alpha}, \end{aligned} \tag{8.4a}$$

and

$$f(k) = \frac{ck}{\sqrt{\Omega(k)}} \sqrt{\frac{\Omega^2(k) - \Omega^2}{2\Omega^2(k) - (ck)^2 - \Omega^2}}, \tag{8.4b}$$

where the electric dipole moment induced in the sample or probe is $\boldsymbol{p}_\alpha$, and its x-, y-, and z-component is $p_{\alpha j}$ ($j = x, y, z$). Note that the electric dipole operator $\hat{\boldsymbol{p}}_\alpha = \boldsymbol{p}_\alpha[\hat{B}(\boldsymbol{r}_\alpha) + \hat{B}^\dagger(\boldsymbol{r}_\alpha)]$ was substituted in (8.1). The absolute value of the wave vector $\boldsymbol{k}$ and the unit vector along the x-, y-, and z-axis are denoted as k and $\boldsymbol{e}_j$, respectively. In (8.4b), c is the speed of light in vacuum, $\Omega(k)$ is the angular frequency proportional to the exciton-polariton energy $\hbar\Omega(k)$, and Ω is the angular frequency proportional to the exciton energy $\hbar\Omega$ in the material of the subsystem (M).

An effective interaction operator $\hat{V}_{\mathrm{eff}}$ in the P-space is derived by including the contribution of the subsystem (M). (The meaning of 'effective' is explained in Sect. D.4.2.) It is expressed as

$$\hat{V}_{\mathrm{eff}} = (P\hat{J}^\dagger\hat{J}P)^{-1/2}(\hat{P}\hat{J}^\dagger\hat{V}\hat{J}P)(P\hat{J}\hat{J}P)^{-1/2}, \tag{8.5}$$

where the projection operator method was applied to $\hat{V}$ of (8.3)[1]. The projection operator P in this equation is given by (D.35) in Appendix D, in terms of $|\phi_1\rangle$ and $|\phi_2\rangle$ as

$$P = |\phi_1\rangle\langle\phi_1| + |\phi_2\rangle\langle\phi_2| \,. \tag{8.6}$$

The operator $\hat{J}$ is given by (D.50) in Appendix D. Its approximate expression is given by (D.78).

In order to derive an explicit functional form of the effective interaction between the sample and probe, two states $|\phi_1\rangle = |s_\mathrm{e}\rangle|p_\mathrm{g}\rangle|0_{(\mathrm{M})}; \boldsymbol{k}, \Omega(k)\rangle$ and $|\phi_2\rangle = |s_\mathrm{g}\rangle|p_\mathrm{e}\rangle|0_{(\mathrm{M})}; \boldsymbol{k}, \Omega(k)\rangle$ of Sect. 8.1 are employed as the initial and final states of the P-space before and after the interaction, respectively. Then, the magnitude of the effective interaction is evaluated from (8.5) as

$$V_\mathrm{eff}(ps) = \langle\phi_2|\hat{V}_\mathrm{eff}|\phi_1\rangle \,. \tag{8.7}$$

Approximating $\hat{J}$ in (8.5) by $\hat{J}^{(1)} = (E_P^0 - E_Q^0)^{-1}Q\hat{V}P$ (see Sect. D.5 for the details), we can rewrite (8.7) in the approximate form as

$$\begin{aligned} V_\mathrm{eff}(ps) &= \langle\phi_2|[P\hat{V}Q(E_P^0 - E_Q^0)^{-1}\hat{V}P + P\hat{V}(E_P^0 - E_Q^0)^{-1}Q\hat{V}P]|\phi_1\rangle \\ &= 2\langle\phi_2|P\hat{V}Q(E_P^0 - E_Q^0)^{-1}\hat{V}P|\phi_1\rangle \\ &= 2\sum_m \langle\phi_2|P\hat{V}Q|m\rangle\langle m|Q(E_P^0 - E_Q^0)^{-1}\hat{V}P|\phi_1\rangle \,, \end{aligned} \tag{8.8}$$

where the operator Q is defined by $1-P$. The quantities E_P^0 and E_Q^0 represent the eigenvalues of the unperturbed Hamiltonian $\hat{H}_0$ in the P- and Q-spaces, respectively. The matrix element $\langle m|Q(E_P^0 - E_Q^0)^{-1}\hat{V}P|\phi_1\rangle$ in (8.8) represents the virtual transition from the initial state $|\phi_1\rangle$ in the P-space to the intermediate state $|m\rangle$ in the Q-space, while the matrix element $\langle\phi_2|P\hat{V}Q|m\rangle$ represents the successive virtual transition from the intermediate state $|m\rangle$ to the final state $|\phi_2\rangle$ in the P-space. To calculate these matrix elements, it should be noted that the operator $\hat{V}$ of (8.3) contains a term composed of $\hat{B}(\boldsymbol{r}_\alpha)$ and $\hat{B}^\dagger(\boldsymbol{r}_\alpha)$, which are operated to the subsystem (N). Further, it contains a term composed of $\hat{\xi}_\lambda(\boldsymbol{k})$ and $\hat{\xi}_\lambda^\dagger(\boldsymbol{k})$, which are operated to the subsystem (M). Then, one can transform (8.8) to[2]

$$V_\mathrm{eff}(ps) = -\frac{1}{(2\pi)^3\epsilon_0}\int \mathrm{d}^3k \left[\frac{K_\mathrm{P}(\boldsymbol{k})K_\mathrm{S}^*(\boldsymbol{k})}{\Omega(k) - \Omega_0(s)} + \frac{K_\mathrm{S}(\boldsymbol{k})K_\mathrm{P}^*(\boldsymbol{k})}{\Omega(k) + \Omega_0(p)}\right] , \tag{8.9}$$

by noting also that $V_\mathrm{eff}(ps)$ takes a nonzero value only when the intermediate state $|m\rangle$ in the Q-space is the state $|1_{(\mathrm{M})}; \boldsymbol{k}, \Omega(k)\rangle$, in which the single exciton–polariton quantum with momentum $\hbar k$ exists in the subsystem (M).

The following three points should be noted for further evaluation of (8.9):

---

[1] See Sect. D.4 for the derivation details

[2] See Sect. D.6 for details

(a) Eigenvalues of the excited energy in the probe and sample are represented by $\Omega_0(p)$ and $\Omega_0(s)$, respectively, which are assumed to be the lowest eigenvalues of the electron energy in the infinitely deep potential wells with widths a_{P} and a_{S}, respectively. These eigenvalues are expressed as

$$\hbar\Omega_0(p) = \frac{3\hbar^2}{2m_{\mathrm{eP}}}\left(\frac{\pi}{a_{\mathrm{P}}}\right)^2 , \tag{8.10a}$$

$$\hbar\Omega_0(s) = \frac{3\hbar^2}{2m_{\mathrm{eS}}}\left(\frac{\pi}{a_{\mathrm{S}}}\right)^2 , \tag{8.10b}$$

respectively, where m_{eP} and m_{eS} are the effective masses of an electron in the sample and probe, respectively.

(b) By approximating a parabolic dispersion relation [the relation between k and $\Omega(k)$] for the exciton–polariton, the energy of the exciton–polariton is expressed as

$$\hbar\Omega(k) = \hbar\Omega + \frac{(\hbar k)^2}{2m_{\mathrm{P}}} , \tag{8.11}$$

where m_{P} represents the effective mass of the exciton–polariton. [See Appendix C for the derivation of (8.11). The effective mass approximation was employed, i.e., the upper branch of the curve in Fig. C.1, which corresponds to the positive term of (C.13), was approximated as a parabolic function of k.]

(c) Since $f(k)$ of (8.4b) can be approximated by a constant, all the quantities in (8.4a) are also treated as constant except for $\exp(\mathrm{i}\boldsymbol{k}\cdot\boldsymbol{r}_\alpha)$.

Following the three points (a)–(c) presented above, the integral of (8.9) can be carried out analytically, and the result is (see Sect. D.7)

$$V_{\mathrm{eff}}(ps) \propto \frac{\exp(-\pi\mu_{\mathrm{P}} r/a_{\mathrm{P}})}{r} + \frac{\exp(\mathrm{i}\pi\mu_{\mathrm{S}} r/a_{\mathrm{S}})}{r} , \tag{8.12}$$

where

$$r = |\boldsymbol{r}_{\mathrm{P}} - \boldsymbol{r}_{\mathrm{S}}| , \quad \mu_{\mathrm{P}} = \frac{\sqrt{3}m_{\mathrm{P}}}{m_{\mathrm{eP}}} , \quad \mu_{\mathrm{S}} = \frac{\sqrt{3}m_{\mathrm{P}}}{m_{\mathrm{eS}}} . \tag{8.13}$$

Here the position vectors $\boldsymbol{r}_{\mathrm{P}}$ and $\boldsymbol{r}_{\mathrm{S}}$ represent arbitrary points in a probe and a sample, respectively. The first term of (8.12) is represented by a Yukawa function with the form $\exp(-\kappa x)/x$, which decays with increasing x, as shown in Fig. 8.2. The decay length can be defined as the position $x = 1/\kappa$ at which $\exp(-\kappa x)$ takes the value e^{-1}. The decay length of the first term of (8.12) is $a_{\mathrm{P}}/\pi\mu_{\mathrm{P}}$, which is proportional to the probe size a_{P}. Equation (8.12) thus shows that there are optical electromagnetic fields around the probe whose extent of spatial distribution is equivalent to the probe size. This is none other than the optical near field, which is localized around the probe like an electron cloud around an atomic nucleus. (Although the optical near field was compared to a thin optical film in Chap. 2, the present discussion shows that it is more appropriate to compare it to an electron cloud around the atomic

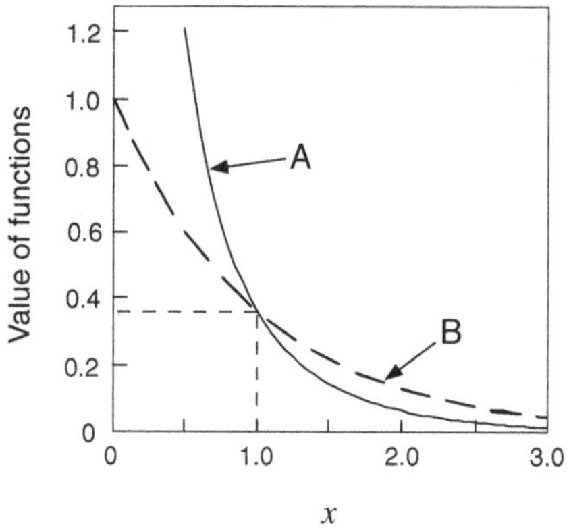

Fig. 8.2. Profiles of a Yukawa function $e^{-\kappa x}/x$ (curve A) and an exponential function $e^{-\kappa x}$ (curve B), where $\kappa = 1$

nucleus. However, it should be noted that the real photon does not form a localized optical field, while the real electron forms a cloud localized around the atomic nucleus.)

Further, this term corresponds to the process in which an exciton–polariton with energy eigenvalue $\hbar\Omega(k)$ emitted from the probe in the ground state is absorbed into the sample in the excited state (see Fig. 8.3a). This is a nonresonant process because it does not follow the energy conservation law. On the other hand, the second term of (8.12) represents conventional propagating light. It corresponds to the resonant process in which an exciton–polariton emitted from the sample in the excited state is absorbed into the probe in the ground state (see Fig. 8.3b), following the energy conservation law. However, it should be noted that both cases represent the virtual transition process mediated by a virtual polariton, which does not follow the energy conservation law. This is well explained by the fact that such a virtual transition process is a quantum mechanical phenomenon, occurring within a sufficiently short period Δt while satisfying the uncertainty principle $\Delta E \Delta t \geq \hbar/2$. The reason why the optical near field, i.e., the localized field, could be successfully derived above was that such a virtual transition process was included in the theoretical model.

In an equivalent way to the discussion from (8.7) to (8.13), the states $|\phi_2\rangle = |s_g\rangle|p_e\rangle|0_{(M)}; \boldsymbol{k}, \Omega(k)\rangle\rangle$ and $|\phi_1\rangle = |s_e\rangle|p_g\rangle|0_{(M)}; \boldsymbol{k}, \Omega(k)\rangle$ of Sect. 8.1 are employed as the initial and final states before and after interaction, respectively. Then the magnitude of interaction $V_{\text{eff}}(sp)$ is evaluated by exchanging the subscripts S and P in (8.8)–(8.12), and is given by

$$V_{\text{eff}}(sp) \propto \frac{\exp(-\pi\mu_S r/a_S)}{r} + \frac{\exp(i\pi\mu_P r/a_P)}{r} . \tag{8.14}$$

The first term of this equation represents the nonresonant process in which an exciton–polariton emitted from the sample in the ground state is absorbed

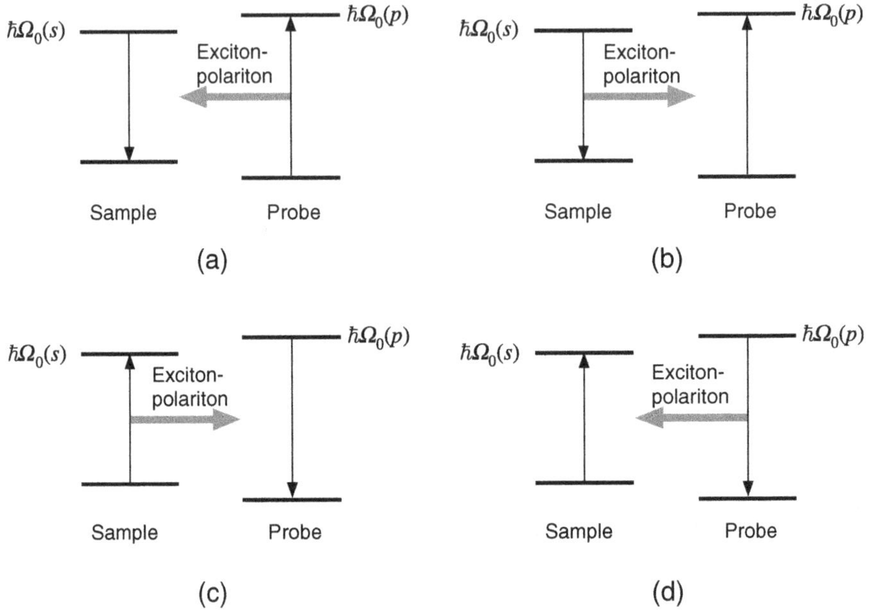

Fig. 8.3. Schematic explanation of energy transfer between the sample and probe. *Thick arrows* represent the absorbed or emitted exciton–polariton. *Thin arrows* represent the transition between energy levels. (**a**) Nonresonant process in which the exciton–polariton is emitted from the probe in the ground state and absorbed into the sample in the excited state. (**b**) Resonant process in which the exciton–polariton is emitted from the sample in the excited state and absorbed into the probe in the ground state. (**c**) Nonresonant process in which the exciton–polariton is emitted from the sample in the ground state and absorbed into the probe in the excited state. (**d**) Resonant process in which the exciton–polariton is emitted from the probe in the excited state and absorbed into the sample in the ground state

into the probe in the excited state (see Fig. 8.3c). The second term is the resonant process in which an exciton–polariton emitted from the probe in the excited state is absorbed into the sample in the ground state (see Fig. 8.3d).

Summing (8.12) and (8.14), the total magnitude of interaction V_{eff} is given by

$$\begin{aligned} V_{\mathrm{eff}} &= V_{\mathrm{eff}}(ps) + V_{\mathrm{eff}}(sp) \\ &\propto \frac{\exp(-\pi\mu_{\mathrm{P}} r/a_{\mathrm{P}})}{r} + \frac{\exp(\mathrm{i}\pi\mu_{\mathrm{S}} r/a_{\mathrm{S}})}{r} \\ &\qquad + \frac{\exp(-\pi\mu_{\mathrm{S}} r/a_{\mathrm{S}})}{r} + \frac{\exp(\mathrm{i}\pi\mu_{\mathrm{P}} r/a_{\mathrm{P}})}{r} \,. \end{aligned} \tag{8.15}$$

The first term of this equation shows that the probe has an optical near field in its vicinity, with decay length equivalent to the probe size. The third term represents that there is an optical near field in the vicinity of the sample.

These two terms are also called Yukawa potentials because they are expressed by a Yukawa function. On the other hand, the second and fourth terms represent that there is a conventional propagating light. Therefore, as the sum of the four terms, (8.15) shows that the optical near field and the propagating scattered light are generated around the two particles, as illustrated in Fig. 2.6 of Chap. 2.

The above discussion can also be applied to a variety of different types of matter, such as gaseous atoms, molecules, quantum dots, and so on, if (8.10a) and (8.10b) are replaced by their energy eigenvalues. The magnitude of interaction can also be expressed by a sum of Yukawa functions. In summary, the present theoretical model, called the Yukawa potential model, can be applied systematically to a variety of materials, ranging from atoms to nanometric materials, in a common theoretical framework.

8.3 Optical Near Field and its Characteristics

Equation (8.15) shows that the electromagnetic interaction between the sample and probe contains a decaying component around them, which is the optical near field, represented by a Yukawa function. Therefore, in the case of a spherical sample or probe with radius a, we may say that a source of such an interaction exists at every position $\boldsymbol{r}$ in the sample and probe. The scalar potential $\phi(\boldsymbol{r})$ of the electromagnetic fields at an arbitrary position $\boldsymbol{r}$ outside the sphere is given by integrating these sources, and is expressed as

$$\phi(\boldsymbol{r}) \propto \int_{\text{sphere}} \frac{\exp(-\mu|\boldsymbol{r}-\boldsymbol{r}'|)}{|\boldsymbol{r}-\boldsymbol{r}'|} \mathrm{d}^3 r' \,. \tag{8.16}$$

From (8.13) and (8.15), μ and μ_α are given by

$$\mu = \frac{\pi\mu_\alpha}{a} \,, \quad \mu_\alpha = \frac{\sqrt{3}m_{\mathrm{P}}}{m_{\mathrm{e}\alpha}} \,, \tag{8.17}$$

respectively, where $m_{\mathrm{e}\alpha}$ is the effective mass of an electron in the sphere. Carrying out the integration in (8.16), one obtains

$$\phi(\boldsymbol{r}) = \frac{2\pi}{\mu^3}\left\{(1+\mu a)\frac{\exp[-\mu(r+a)]}{r} - (1-\mu a)\frac{\exp[-\mu(r-a)]}{r}\right\} \,, \tag{8.18}$$

for $r = |\boldsymbol{r}| > a$. The right-hand side is given by the difference of two Yukawa functions. (Refer to Problem 8.1 at the end of this chapter for the derivation.) The decay length r_{n} of the optical near field can be defined in the same way as the decay length of (8.12) and is expressed by (8.17) and (8.18) as

$$r_{\mathrm{n}} = \frac{1}{\mu} = \frac{a}{\pi\mu_\alpha} \,, \tag{8.19}$$

which corresponds to the size of the sphere.

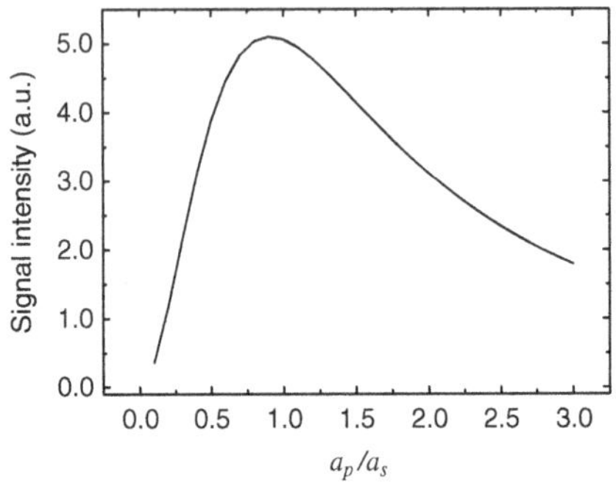

Fig. 8.4. Calculated results for the relation between the ratio a_P/a_S and the signal intensity $I(a_P, a_S, R, z)$ detected while scanning the sphere P at $z = 1\,\mathrm{nm}$

Let us give several examples as an application of the present theory. First, Fig. 4.4 shows the simplest model of a near-field optical microscope. The sample sphere S (radius a_S) is fixed at the origin of the coordinate system, while the height of the probe sphere P (radius a_P) is maintained at z from the top of the sphere S during scanning, unlike the case in Fig. 4.4. The signal intensity at position $\boldsymbol{r}$ in the sphere P is derived from (8.18), and its integral over the sphere P corresponds to the detectable signal intensity $I(a_P, a_S, R, z)$, where R is the distance between the two sphere centers (i.e., the centers of P and S). Therefore, the signal intensity can be expressed by the Yukawa function, as in (8.18). Figure 8.4 shows the calculated result of the relation between the ratio a_P/a_S and the signal intensity detected while scanning the sphere P at $z = 1\,\mathrm{nm}$. The signal intensity reaches its maximum at $a_P = a_S$, which corresponds to the size-dependent resonance given in Fig. 4.8.

Second, the system shown in Fig. 4.15 is employed in order to express the profile of a tapered fiber probe in an approximate manner. The radius of the sphere T in the tapered part is given by (4.19), which depends on the cone angle θ. Assume that the probe is scanning whilst maintaining its height z, as in Fig. 8.4. Figures 8.5a and b show the calculated signal intensities for $\theta = 80°$ and $20°$, respectively, as a function of the position of the sphere P along the x-axis, which correspond to Fig. 4.16. Comparing Figs. 8.5a and b, it is found that the signal intensity detected by the sphere T is larger than that detected by the sphere P for large θ. Curve C is therefore broader in Fig. 8.5a than in Fig. 8.5b. This means that the visibility defined in Sect. 4.3.1 is low. The magnitude of the interaction between the spheres T and S becomes lower for smaller θ. Therefore, the signal intensity is mainly contributed from the sphere P. In consequence, higher visibility can be obtained, as shown in Fig. 8.5b. This is attributed to the facts that the optical near-field interaction is effective only within the spatial range of the sphere sizes and that the

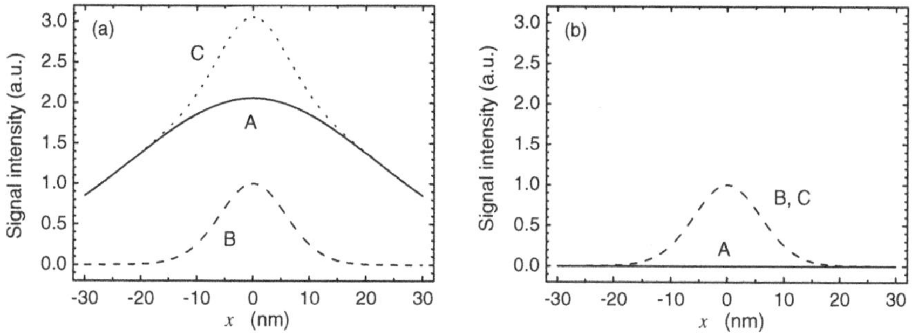

Fig. 8.5. Calculated signal intensities for (**a**) $\theta = 80°$ and (**b**) $20°$ as a function of the position of the sphere P along the x-axis. Curves A and B represent the contributions from spheres T and P, respectively. Curve C is the total intensity

magnitude of interaction decreases rapidly with increasing distance between the spheres.

The visibility also depends on the height z. If z becomes larger than the effective range of interaction between the spheres P and S, the signal intensity is mainly contributed by the interaction between the spheres T and S. In such a case, the sphere T is the main contributor to the signal intensity because the radius of the sphere T is larger than that of the sphere S. It therefore leads to lower visibility. Hence, the values of θ and z must be maintained sufficiently small in order to achieve higher visibility. In order to improve the visibility, it is more effective to coat the tapered part with an opaque film, since it may eliminate the effect of the tapered part.

Finally, the dependence of the signal intensity on the polarization state of the incident light is discussed by noting the numerator of (D.87) in Appendix D, which is given by

$$\sum_{\lambda=1}^{2}\sum_{i=1}^{3}\sum_{j=1}^{3} p_{\mathrm{S}j}\left[\boldsymbol{e}_j\cdot\boldsymbol{e}_\lambda(\boldsymbol{k})\right] p_{\mathrm{P}i}\left[\boldsymbol{p}\cdot\boldsymbol{e}_\lambda(\boldsymbol{k})\right]\mathrm{e}^{\mathrm{i}\boldsymbol{k}\cdot(\boldsymbol{r}_\mathrm{S}-\boldsymbol{r}_\mathrm{P})} \tag{8.20}$$

$$= \left[\boldsymbol{p}_\mathrm{S}\cdot\boldsymbol{p}_\mathrm{P} + \frac{(\boldsymbol{p}_\mathrm{S}\cdot\nabla)(\boldsymbol{p}_\mathrm{P}\cdot\nabla)}{k^2}\right]\mathrm{e}^{\mathrm{i}\boldsymbol{k}\cdot(\boldsymbol{r}_\mathrm{S}-\boldsymbol{r}_\mathrm{P})}\,.$$

The second term on the right-hand side contains a spatial differential operator ∇, which is responsible for the dependence on the polarization state of the incident light. Although this term was neglected for simplicity in deriving V_{eff} in (8.12), (8.14), and (8.15), it must be considered for the present numerical calculation of the spatial distribution of the signal intensity.

Figure 8.6a shows a schematic configuration for the calculation, where the sphere P (radius $a_\mathrm{P} = 10\,\mathrm{nm}$) scans two-dimensionally on a circular aperture ($a_\mathrm{S} = 10\,\mathrm{nm}$) at height $z = 1\,\mathrm{nm}$. Figure 8.6b shows the calculated result when the incident light is polarized along the x-axis. The x- and y-axes represent the position of the sphere P. This figure clearly shows that

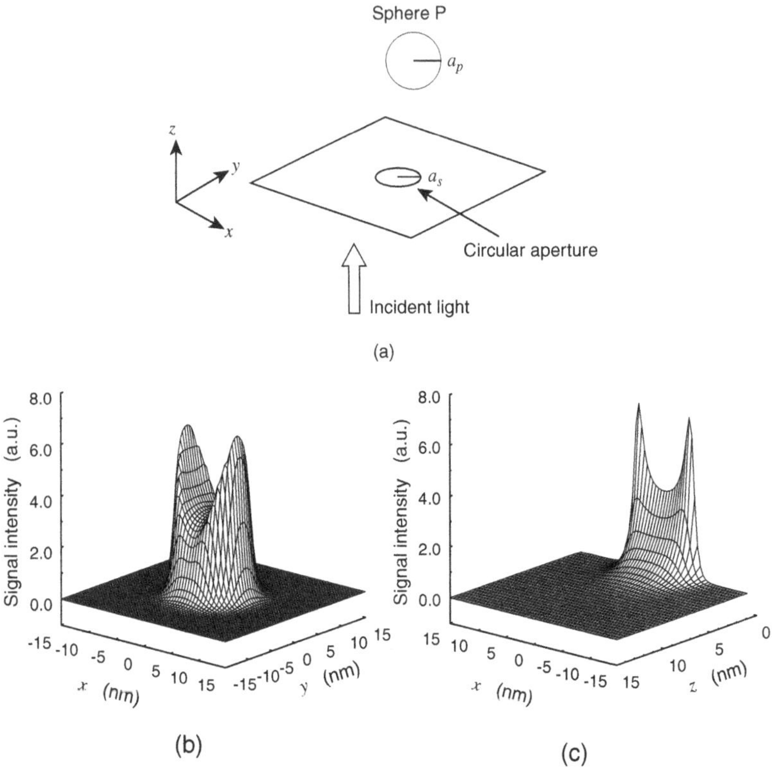

Fig. 8.6. (**a**) Schematic configuration for the calculation, where sphere P (radius $a_{\mathrm{P}} = 10\,\mathrm{nm}$) scans two-dimensionally on a circular aperture ($a_{\mathrm{S}} = 10\,\mathrm{nm}$) whilst maintaining the height $z = 1\,\mathrm{nm}$. (**b**) Calculated result when the incident light is polarized along the x-axis. Values on the x- and y-axes represent the position of sphere P. (**c**) Calculated result when sphere P scans in the xz-plane at $y = 0$

the spatial distribution of the signal intensity has two peaks at the edges of the aperture, which correspond to the edge effect occurring in the case of s-polarized incident light, as described in Sect. 4.2.4. Figure 8.6c shows the calculated result when the sphere P scans in the xz-plane at $y = 0$. The rapid decay of the signal intensity can be clearly seen with increasing distance between the sphere P and the aperture. The decay length is several nanometers.

Problems

Problem 8.1

Starting from (8.16), derive (8.18).

9 Application to Nanophotonics and Atom Photonics

Utilizing the theoretical basis presented in Chaps. 4–8, the present chapter discusses the possibility of creating new fields in nanophotonics and atom photonics, which shift the paradigm of optical science and technology.

9.1 Energy Transfer Between Molecules and Application to Optical Near-Field Measurement

One important aim in molecular spectroscopy is to obtain high spatial resolution images of the light emitted from one molecule using an adjacent molecule as an illuminating light source. For this purpose, numerous experiments have been carried out over the past half century. The ultimate goal is to discriminate molecules with resolution down to molecular sizes. The present section reviews optical-near-field technology aiming to achieve the goal.

9.1.1 Radiative Energy Transfer

Consider two dye molecules close to each other, as shown in Fig. 9.1a. The frequency dependence of their absorption and emission spectra is shown in Fig. 9.1b, in which ν_{da}, ν_{de}, ν_{aa}, and ν_{ae} represent the center frequencies of absorption and emission spectra of the first and second dye molecules, known as acceptor and donor molecules, respectively. It is assumed that they satisfy the relation $\nu_{\mathrm{ae}} < \nu_{\mathrm{aa}} \approx \nu_{\mathrm{de}} < \nu_{\mathrm{da}}$, i.e., the emission spectrum of the donor overlaps with the absorption spectrum of the acceptor on the frequency axis.

If the two dye molecules are illuminated by light with frequency ν_{ex}, and $\nu_{\mathrm{da}} < \nu_{\mathrm{ex}}$, the donor will be excited by absorbing the light. After the absorption, energy is transferred from the excited donor to the acceptor due to the dipole–dipole interaction reviewed in Chap. 4. This phenomenon is called energy transfer [9.1–9.3]. Such energy transfer is possible when $\nu_{\mathrm{aa}} \approx \nu_{\mathrm{de}}$. Furthermore, when the acceptor radiates light involving the excitation by energy transfer, such a phenomenon is called radiative energy transfer. Radiative energy transfer is applicable to optical-near-field measurement when the distance between two molecules is within the Förster dipole–dipole energy transfer radius R_0, i.e., several nanometers [9.3].

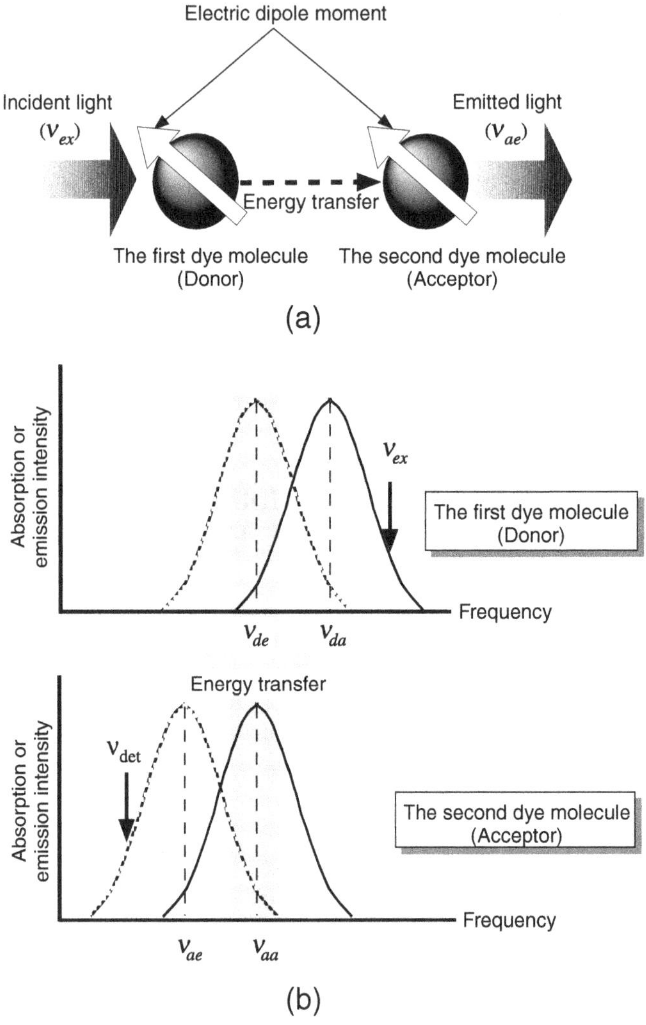

Fig. 9.1. Energy transfer and its application. (**a**) Principle of energy transfer. (**b**) Relation between the absorption and emission spectra of two dye molecules, where ν_{ex} and ν_{det} are the frequencies of the incident light and detected component of the emission from the acceptor, respectively

Figure 9.2 shows an example of such measurements, in which an acceptor molecule is installed on the apex of a fiber probe for collection-mode near-field optical measurement. The molecule emits the light via radiative energy transfer from the donor molecule, and its frequency component ν_{det} is selectively detected using a band-pass filter. When $\nu_{det} < \nu_{ae}$, the incident light and the emitted light from the donor cannot pass through the filter, which means that the contrast C can be dramatically enhanced by excluding back-

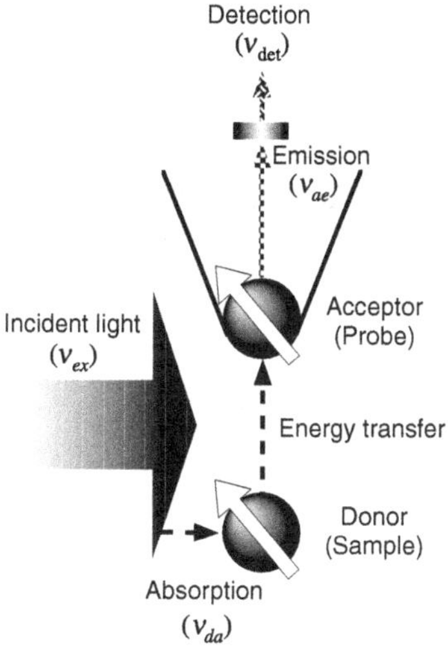

Fig. 9.2. Application of radiative energy transfer to collection-mode near-field optical microscopy

ground signals, as shown in Sect. 4.2.3. Since emission from the acceptor is detected only when the sample–probe separation is less than R_0, the image of the donor molecule is obtained with high resolution. The tapered part of the fiber probe does not give any contributions because the acceptor is fixed only on its apex, which results in the dramatic increase in the visibility V of Sect. 4.3.1.

The measurement described above is also applicable in the illumination mode, i.e., by installing a donor dye molecule at the apex of the fiber probe.

9.1.2 Non-Radiative Energy Transfer

Frequency conversion is used for the measurements mentioned in Sect. 9.1.1, using the energy transfer between the donor and acceptor. However, it is not technically straightforward to install an acceptor molecule at the apex of the fiber probe. As an easy technique, it is advantageous to use non-radiative energy transfer by installing a metal particle at the apex of the fiber probe. Energy is transferred from the excited donor to the metal particle by the principle shown in Fig. 9.1. However, the metal dissipates the transferred energy thermally instead of emitting light. This is why this energy transfer is called non-radiative energy transfer. Therefore, when a probe is scanned over a donor, the center of the image becomes dark, i.e., we have an inverted image. High resolution, down to several nanometers, can also be obtained in this imaging, as was the case using radiative energy transfer (Fig. 9.2). Comparing with radiative energy transfer, non-radiative energy transfer is easy to measure because the metal-coated fiber probe in Fig. 3.1e is used for the measurement, and this can be reproducibly fabricated.

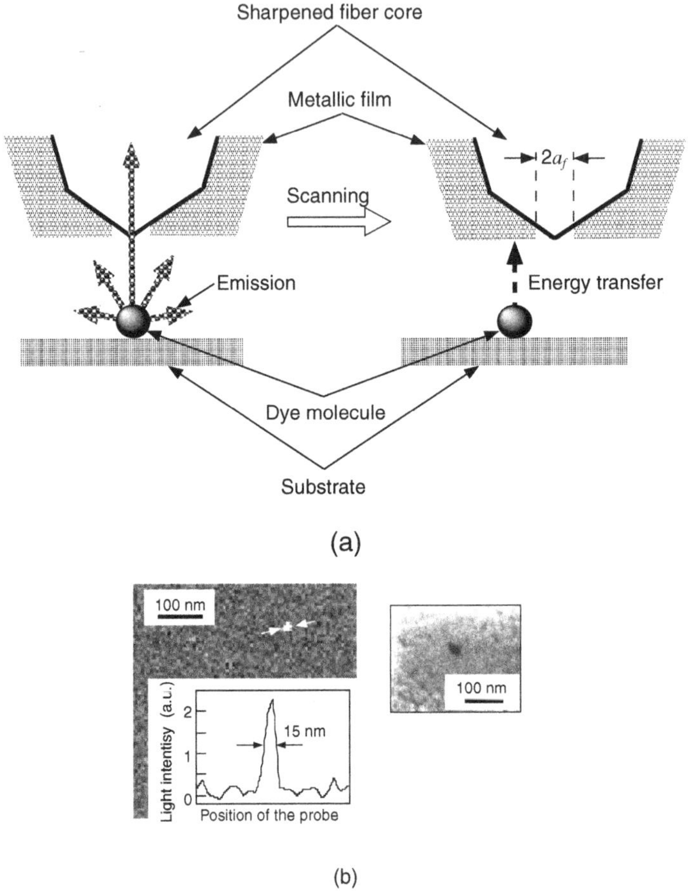

Fig. 9.3. Application of non-radiative energy transfer to collection-mode near-field optical microscopy. (**a**) Explanation of the principle. *Left*: emission from the dye molecule when the tip of a sharpened fiber core is located right over the dye molecule. *Right*: non-radiative energy transfer takes place when the metallic film coated on the fiber probe is located right over the dye molecule. (**b**) Experimental results. *Right*: SEM image of the fiber probe tip used for the experiment, equivalent to the left-hand figure of Fig. 3.1d. The diameter of the *black circle* at the center corresponds to $2a_f$ (= 20 nm). *Left*: spatial distribution of the light intensity emitted from a single Cy5.5 dye molecule, whose cross-sectional profile along the line indicated by two *white arrows* is represented by the lower curve

A method for obtaining non-inverted images has been demonstrated using the fiber probe shown in Fig. 3.1d, in which the tip of the sharpened fiber core is buried into the metallic film. When the fiber probe is located right over the dye molecule (see the left-hand part of Fig. 9.3a), the dye molecule is excited by the incident light and emits light which is efficiently detected through the fiber probe. When the fiber probe is moved horizontally (see the right-hand part of Fig. 9.3a), non-radiative energy transfer takes place from the dye molecule (donor) to the metallic film (acceptor) coated on the fiber probe, and light emission is prohibited. A non-inverted image can thus be obtained by scanning the metal-coated fiber probe. The spatial distribution of the emitted light intensity obtained by scanning is governed by the diameter $2a_f$ of the central hole on the metallic top surface. Thus, a higher-resolution image of the dye molecule is expected using a fiber probe with smaller a_f.

The left-hand image in Fig. 9.3b shows an example of such a non-inverted image of the light intensity emitted from a single Cy5.5 dye molecule [9.4]. The fiber probe used here is the same as in Fig. 3.1d, where $a_f = 10$ nm (see the SEM picture on the right in Fig. 9.3b). The diameter of the image is about 15 nm (see the cross-sectional profile shown in the inset of Fig. 9.3b), which is comparable to the diameter $2a_f$ (= 20 nm). However, it should be noted that the value of a_f was measured through the SEM image, as pointed out in relation to Fig. 3.1d. The light can transmit through a very thin metallic film around the rim of the central hole on the metallic top surface. Therefore, the optically effective diameter $2a_f$ is larger than 20 nm. The reason why the image diameter is as small as 15 nm is attributed to the non-radiative energy transfer taking place around the rim of the central hole. Thus, the image diameter of 15 nm is much smaller than the optically effective diameter of the central hole. A recent experiment has obtained an image with diameter as small as 8 nm, using the same fiber probe as shown in Fig. 3.1d [9.5].

As an example similar to the measurement shown in Fig. 9.3a, the dependence of the emission intensity and lifetime has been measured as a function of the position of the fiber probe [9.6].

9.2 Atom Manipulation

Following the review on the development of atom photonics in Sect. 3.3.5, this section describes the basic concepts in terms of the theories of Chaps. 4–8. Concepts are formulated for controlling and manipulating the thermal motion of a neutral atom in vacuum by means of an optical near field. Section 9.2.1 presents the formulation based on the conventional theory [9.7]. It should be noted that it is an approximate formulation which becomes less accurate when the material size is decreased to as small as several nanometers. In order to overcome this difficulty, Sect. 9.2.2 provides a rigorous formulation based on the theory in Chap. 8.

9.2.1 Formulation by Conventional Theory

Dipole Force

When an atom absorbs light whose frequency corresponds to the energy difference $E_u - E_l$ between the two atomic energy levels (E_u and E_l are the

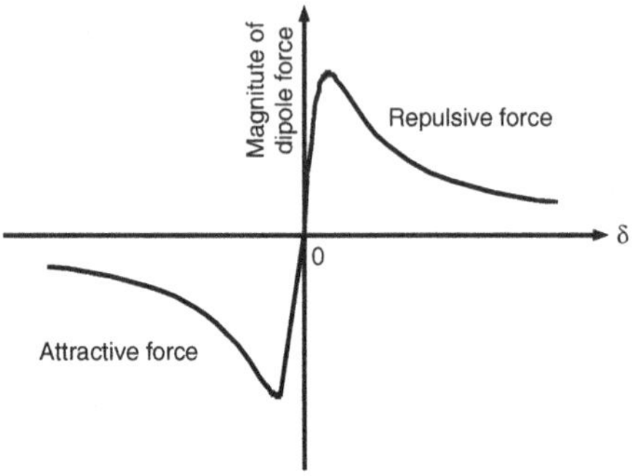

Fig. 9.4. Relation between the detuning δ and the magnitude of the dipole force

energies of the upper and lower levels, respectively), an electric dipole is induced in the atom. The dipole force $\boldsymbol{F}_{\mathrm{d}}$ is generated when the gradient of the electric field of the light acts on the electric dipole. If the distribution of the light intensity is spatially inhomogeneous, the atom is driven by the dipole force.[1]

The direction of the dipole force $\boldsymbol{F}_{\mathrm{d}}$ depends on the detuning δ $(= \omega_{\mathrm{L}} - \omega_0)$, i.e., the difference between the angular frequency ω_{L} of light and the atomic resonant angular frequency $\omega_0 = (E_{\mathrm{u}} - E_{\mathrm{l}})/\hbar$.[2] Here, a two-energy-level approximation is used for simplicity. This is valid for the alkali atoms such as Rb given in Sect. 3.3.5. If the electric field $\boldsymbol{E}$ of the light induces the electric dipole moment $\boldsymbol{p}$ in the two-energy-level atom, the dipole force on the atom is given by

$$\boldsymbol{F}_{\mathrm{d}} = -\frac{\hbar\delta\nabla\Omega^2}{4\delta^2 + \gamma^2 + 2\Omega^2}, \tag{9.1}$$

where ∇ represents the differential operator given in Sect. A.1 and $\Omega = pE/\hbar$ is called the Rabi angular frequency of the atom. The relaxation constant γ is inversely proportional to the time constant τ of the relaxation from the upper energy level of the atom, i.e., $\gamma = 2\pi/\tau$.

Figure 9.4 shows the magnitude of $\boldsymbol{F}_{\mathrm{d}}$ given by (9.1). It follows from (9.1) that $\boldsymbol{F}_{\mathrm{d}}$ is parallel to the direction of decreasing light intensity when $\delta > 0$, i.e., it is a repulsive force, so that the atom would be flicked out of the optical field. In the case of $\delta < 0$, on the other hand, $\boldsymbol{F}_{\mathrm{d}}$ is parallel to the direction of increasing light intensity, i.e., it is an attractive force, so that the atom would be drawn into the optical field. It should be noted that the curve in Fig. 3.19a corresponds to that of Fig. 9.4.

[1] This force is also called a gradient force because its magnitude is proportional to the gradient of the light intensity.

[2] To simplify the notation, an angular frequency is used here instead of the frequency. Since the angular frequency is 2π times the frequency, the atomic resonance frequency ν_0 is expressed as $\omega_0/2\pi = (E_{\mathrm{u}} - E_{\mathrm{l}})/h$.

Atom Reflection Using the Optical Near Field on a Planar Surface

An atom in vacuum can be reflected by the repulsive dipole force induced by an optical near field on a planar surface of a dielectric material such as a glass, as explained schematically in Fig. 9.5. The optical near field on a planar surface is also called evanescent light, as described in Sect. 2.1. Its intensity decreases with increasing distance from the planar surface. Therefore, if an atom approaches the dielectric surface, it is reflected by the optical near field with $\delta > 0$, i.e., the dielectric surface works as a kind of atomic mirror to reflect the atom.

The intensity and wavelength of the incident light on a material surface and the refractive index of the dielectric are expressed by I_0, λ, and n_{m}, respectively. Total reflection takes place if the incidence angle θ_1 is larger than the critical angle θ_{c} of (2.2). Since (2.6) represents the electric field amplitude of the optical near field generated under this condition, the optical-near-field intensity can be obtained as a function of the distance r from the dielectric surface. It is expressed as $I(r) = I_0 \exp(-2r/\Lambda)$, where $\Lambda = (\lambda/2\pi)\sqrt{n^2 \sin^2 \theta_1 - 1}$ is the decay length of the electric field amplitude given by (2.7). The absolute value of $\boldsymbol{F}_{\mathrm{d}}$ in (9.1) is then given by

$$\boldsymbol{F}_{\mathrm{d}} = \frac{2\hbar \Delta \Omega^2(r)/\Lambda}{4\Delta^2 + \gamma^2 + 2\Omega^2(r)} , \tag{9.2}$$

where $\Delta = \omega_{\mathrm{L}} - \Omega_0 - k_{\mathrm{t}}\nu_z$ is the detuning, which includes the Doppler effect due to the thermal velocity component of an atom ν_z along the dielectric surface, and $k_{\mathrm{t}} = n(\omega_{\mathrm{L}}/c)\sin\theta_1$ is the wave number. The Rabi angular frequency $\Omega(r)$ depends on the distance r, which is defined by $\Omega(r) = \gamma\sqrt{I(r)/(2I_{\mathrm{s}})}$, where $I_{\mathrm{s}} = \hbar\gamma\omega_{\mathrm{L}}^3/(12\pi c^2)$ is called the saturation intensity for the electric dipole transition.

The first experiment on atom reflection using the optical near field was successfully carried out with Na atoms [9.8]. This kind of experiment has re-

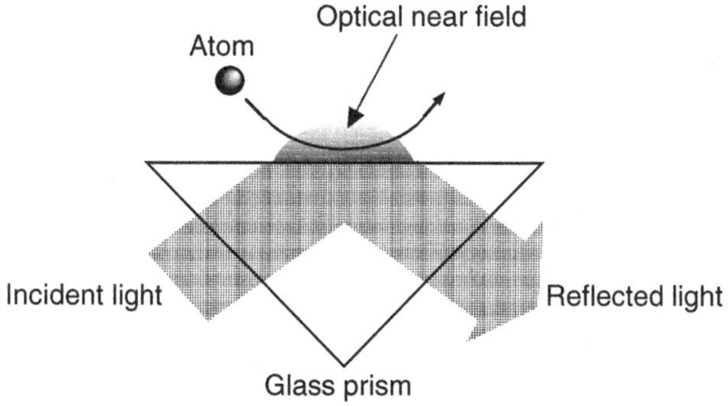

Fig. 9.5. Atom deflection by the optical near field on a planar surface

cently been utilized to observe the van der Waals interaction between ground state Cs atoms and a dielectric material surface [9.9].

Atom Guidance Using the Optical Near Field on the Inner Wall of a Hollow Optical Fiber

When using an optical near field on a planar dielectric surface, the in-surface component of the trajectories of the reflected atoms cannot be controlled, i.e., the degree of freedom of the residual thermal motion of the reflected atoms is two-dimensional. In contrast, if atoms are guided through a hollow optical fiber using the optical near field generated on the inner surface, the atoms show one-dimensional movement, as was described in Sect. 3.3.5. Thus the controllability of the trajectory of the transmitted atom can be improved. This atom guidance is formulated in the following.

The modes of light guided through the core of a hollow optical fiber depend on the diameters and refractive indices of the core and cladding, and the wavelength of the light. Assume that Rb atoms are guided through the hollow optical fiber shown in Fig. 3.18. The angular frequency of the light (wavelength $\lambda = 780\,\mathrm{nm}$) is close to that of the atomic resonance ω_0. It has been found that there exists a guided mode named LP_{m1} mode ($m = 0, 1, 2, \ldots$) for this optical fiber.[3] The cross-sectional spatial distribution of the light intensity of the LP_{m1} mode is expressed using a polar coordinate (ρ, ϕ) as [9.10]

$$I(\rho,\phi) = \begin{cases} \alpha B_1^2 I_m^2(v\rho)\cos^2(m\phi+\psi) & \rho < a\,, \\ \alpha\left[B_2 J_m(u\rho) + B_3 N_m(u\rho)\right]^2 \cos^2(m\phi+\psi) & a \le \rho \le a+d\,, \\ \alpha B_4^2 K_m^2(w\rho)\cos^2(m\phi+\psi) & (a+d<\rho)\,, \end{cases} \tag{9.3}$$

where a is the hollow radius, d the thickness of the doughnut-shaped core, m an integer, and ψ an arbitrary phase constant. The constant α is defined as $\beta/\omega_{\mathrm{L}}\mu_0$, β is the propagation constant, and B_i ($i = 1, \ldots, 4$) are the constants to be determined by boundary conditions at $\rho = a$ and $\rho = a + d$. Functions $J_m(u\rho)$, $N_m(u\rho)$, $I_m(v\rho)$, and $K_m(w\rho)$ are the m th order Bessel and modified Bessel functions of the first and second kinds, respectively. Assuming that the refractive indices of the core and cladding are n_1 and n_2, respectively, three transverse characteristic constants are expressed as $u = \sqrt{n_1^2k^2 - \beta^2}$, $v = \sqrt{\beta^2 - k^2}$, and $w = \sqrt{\beta^2 - n_2^2k^2}$. Equation (9.3) shows that the cross-sectional spatial distribution of the light intensity of the LP_{01} ($m = 0$), LP_{11} ($m = 1$), and LP_{21} ($m = 2$) modes are doughnut-shaped without any nodes,

---

[3] Exact analysis has found that there exist several guided modes named TE_{01}, TM_{01}, HE_{11}, HE_{21}, HE_{31}, and EH_{11}. However, this chapter employs the weak guiding approximation (WGA) for simpler analysis [9.10]. The LP_{m1} mode is obtained under this approximation.

butterfly-shaped with two nodes (at $\phi = 0$ and π), and four-leaved-clover-shaped with four nodes (at $\phi = 0$, $\pi/2$, π, and $3\pi/2$), respectively.

The optical-near-field intensity on the inner wall surface of the hollow optical fiber is given by the first line on the right-hand side of (9.3):

$$I_{\mathrm{nf}}(\rho, \phi) = \frac{\beta}{2\omega_{\mathrm{L}}\mu_0} B_1^2 I_m^2(v\rho) \cos^2(m\phi) \,, \tag{9.4}$$

where ψ is fixed at 0 for simplicity. For a large detuning δ, i.e., for the angular frequency of the light ω_{L} that is sufficiently detuned from that of the atomic resonance ω_0, the spontaneous emission rate from the atom is so low that alkaline atoms such as Rb can be regarded as two-level atoms with sufficient accuracy. Under this approximation, the optical potential for reflecting atoms against the inner wall surface is given by

$$U_{\mathrm{opt}}(\rho, \phi) = \frac{\hbar\Delta}{2} \ln\left[1 + \frac{I_{\mathrm{nf}}(\rho, \phi)}{I_{\mathrm{s}}} \frac{\gamma^2}{4\Delta^2 + \gamma^2}\right] \,, \tag{9.5}$$

where $\Delta = \omega_{\mathrm{L}} - \omega_0 - \beta v_z$ is the detuning including the magnitude of the Doppler effect due to the fiber-axis component v_z of the thermal velocity of the atom. In the case of the D_2 spectral line of the Rb atom, typical numerical values of I_{s} and γ are $1.6\,\mathrm{mW/cm^2}$ and $2\pi \times 6.1\,\mathrm{MHz}$, respectively. The magnitude of the optical potential due to the LP_{m1} mode is obtained by substituting (9.4) into (9.5). The LP_{01} mode is used for atom guidance because its doughnut-shaped profile without any nodes can confine the atom in a stable manner.

Stable atom guidance is disturbed if the atoms are attracted and adsorbed onto the inner wall surface of the hollow optical fiber due to the attractive van der Waals force. The effect of adsorption is described as a cavity quantum electrodynamic (QED) effect when an atom exists in a small cavity [9.11]. The cavity QED effect originates from the interaction between an atom and the vacuum modified by the cavity [9.12], which yields two kinds of attractive force: the van der Waals force (see Fig. 5.4) and the Casimir–Polder force. The latter gives rise to a position-dependent Lamb shift of the atomic energy levels.

In spite of the importance for a deeper understanding of the nature of the vacuum, few experimental results on the cavity QED effect have been reported except for some simple cases for a plane or a planar cavity. The difficulty lies in the fact that the cavity QED effect is so small that special delicacy is required for the measurement. However, this effect has to be discussed because it is important for atom manipulation by the optical near field. The cavity QED effect gives rise to an energy shift in an atomic state (see Fig. 5.4). In general, the formulae for the atomic energy shifts in the case of curved dielectric surfaces are very complicated, so that their analysis requires considerable computation time. Here, for simplicity, we adopt an approximate method using a formula for a planar conductor cavity in order to analyze the case of the inner wall surface of a hollow optical fiber.

The cavity potential $U_{\text{cav}}(\rho)$ on an atom inside a planar conductor cavity with a space interval $2a$ is written in a simple analytical form as [9.11]

$$U_{\text{cav}}(\rho) = -\sum_e \frac{\pi |d_{eg}|^2}{48\varepsilon_0 a^3} \int_0^\infty \frac{r^2 \cosh(\pi r \rho/a)}{\sinh(\pi r)} \tan^{-1}\left(\frac{r\lambda_{eg}}{4a}\right) \mathrm{d}r \;, \tag{9.6}$$

where d_{eg} and λ_{eg} are matrix elements of electric dipole transitions from a ground state and wavelengths corresponding to the transitions, respectively. The symbol $\sum_e$ represents summation over all relevant excited states. Consider an atom guided through a hollow optical fiber with hollow radius a by a blue-detuned ($\delta > 0$) optical near field with LP_{01} mode. Then, the total potential $U_{\text{t}}(\rho)$ is given by summing the optical potential $U_{\text{opt}}(\rho)$ and the modified cavity potential $fU_{\text{cav}}(\rho)$: $U_{\text{t}}(\rho) = U_{\text{opt}}(\rho) + fU_{\text{cav}}(\rho)$, where a scaling coefficient f is introduced to approximate the dielectric cylinder case. In addition, it is assumed that the scaling coefficient f can be the product of a dielectric factor ξ and a geometric factor η. Moreover, since the inner wall is considered to be approximately planar for an atom near the surface, the dielectric factor ξ can be written as $\xi = (n_1^2 - 1)/(n_1^2 + 1)$, where n_1 is the refractive index of the core.[4] The geometric factor η is used to convert from a planar to a curved surface. The magnitude of the cavity potential increases with decreasing hollow radius.

Figure 9.6 shows a potential barrier, plotted as a function of the distance $R\ (= a - \rho)$ from the inner wall surface, for reflecting ^{85}Rb atoms. The solid curve shows the total potential $U_{\text{t}}(\rho)$, which is composed of a cavity potential $U_{\text{cav}}(\rho)$ with $f = 1$ and an optical potential $U_{\text{opt}}(\rho)$ under excitation of the LP_{01} mode with a power of 100 mW and a blue detuning of +1 GHz. The broken curve shows an optical potential $U_{\text{opt}}(\rho)$. Comparison between the two curves shows that the potential barrier is greatly reduced due to the cavity potential near the surface. Therefore, if the trajectory of the atomic motion is not parallel to the fiber axis, the atom may jump over the potential barrier and be adsorbed on the inner wall surface. In order to avoid this kind of phenomenon, the trajectory of the atomic motion must be carefully aligned with the fiber axis.

Since the magnitude of the optical potential is proportional to the optical power, it is lower than the magnitude of the cavity potential, i.e., $U_{\text{opt}}(\rho) < fU_{\text{cav}}(\rho)$ for a low optical power, and atoms are adsorbed on the inner wall surface. This means that there exists a threshold optical power for atom guidance. The black arrow in the inset of Fig. 3.19b indicates the threshold optical power. The total potential $U_{\text{t}}(\rho)$ is zero at this threshold, and from this condition, the scaling factor f is derived as $f = |U_{\text{opt}}(\rho)/U_{\text{cav}}(\rho)|$. Equations (9.4)–(9.6) and the black arrow in the inset of Fig. 3.19b lead to the estimate $f \approx 1.3$, which implies a geometric factor $\eta \approx 3.7$. This is a

[4] This equation corresponds to $(\varepsilon - \varepsilon_0)/(\varepsilon + \varepsilon_0)$ in (Q5.7) for the solution to Problem 5.1.

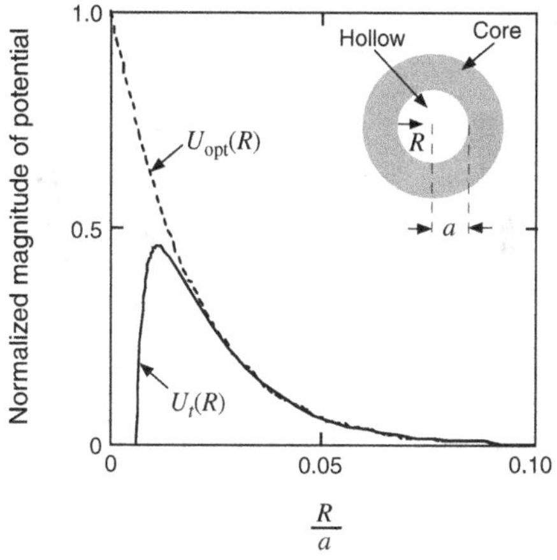

Fig. 9.6. Relation between the distance $R = a - \rho$ from the inner wall surface of a hollow fiber and the value of the potential for an ^{85}Rb atom. *Solid* and *broken curves* represent the total potential $U_t(R)$ and optical potential $U_{opt}(R)$, respectively. Values on the *vertical axis* are normalized to the value of $U_{opt}(0)$

reasonable value because qualitative theoretical analysis has shown that the geometrical factor should be between 1 and 10.

Two kinds of force contribute to the cavity potential of (9.6). The van der Waals force is dominant in the sub-wavelength region, but the Casimir–Polder force is dominant in the super-wavelength region. It follows that the former acts in close proximity to the material surface, while the latter acts in the far-field region. Since the wavelength of light for guiding Rb atoms is 780 nm, a hollow optical fiber with $a = 300$ nm and 1.4 μm corresponds to the van der Waals and Casimir–Polder cases, respectively.

9.2.2 Deflecting and Trapping an Atom Using the Optical Near Field Generated at a Fiber Probe Tip

Higher controllability of the thermal motion of an atom can be obtained by using the optical near field on the inner wall of a hollow optical fiber (the third method described in Sect. 9.2.1) than by using the optical near field on a planar surface (the second method described in Sect. 9.2.1). This is because in the third method the degree of freedom in the thermal motion of an atom is one-dimensional, whereas it is two-dimensional in the second method. However, it should be noted that, even in the third method, the decay length of the optical near field is of the order of the light wavelength, as described by (9.4).

For further accuracy in controlling the thermal motion of an atom, we now discuss the deflection and trapping of an atom, as explained schematically in Figs. 9.7a and b, respectively. For the illumination-mode near-field

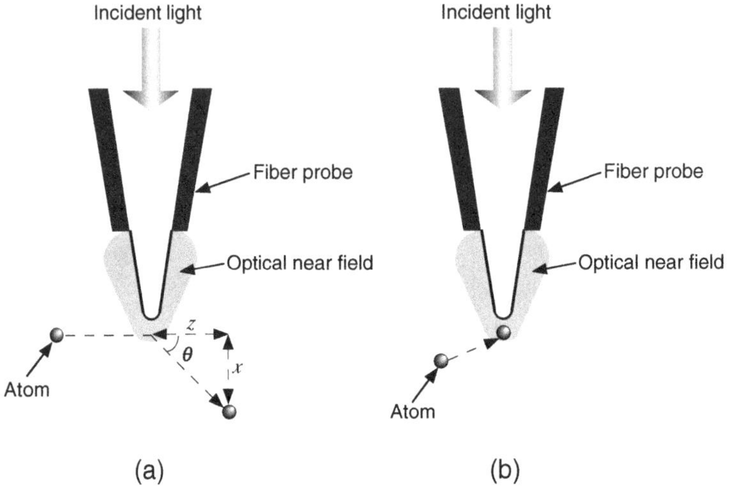

Fig. 9.7. Atom manipulation by the optical near field generated on a fiber probe tip. (**a**) Atom deflection. θ is the deflection angle and x is the displacement due to deflection at position z downstream from the fiber probe. (**b**) Atom trapping

optical microscope, deflection and trapping of an atom are accomplished by the dipole force due to the optical near field generated by a fiber probe tip. In particular, a dramatic increase in accuracy is expected in the case shown in Fig. 9.7b, because the degree of freedom in the residual thermal motion of an atom is zero-dimensional due to the nanometric confinement effect imposed by the optical near field.

However, the three theoretical models described in Sect. 9.2.1 are not accurate enough to describe the mechanical interactions between atoms and the optical near field localized in a space of sub-wavelength dimensions.[5] Instead of using these theoretical models, we adopt the one described in Chap. 8, i.e., the Yukawa potential model, to overcome this difficulty. Using the definition of μ_l and μ_h of (D.88) in Appendix D, two kinds of effective masses are defined, viz.,

$$\Delta_{\mathrm{P}\pm} = \sqrt{\frac{2m_\mathrm{P}[\Omega \pm \Omega_0(p)]}{\hbar}}\,, \quad \Delta_{\mathrm{S}\pm} = \sqrt{\frac{2m_\mathrm{P}[\Omega \pm \Omega_0(s)]}{\hbar}}\,, \tag{9.7}$$

[5] The discussion in previous sections employed Ehrenfest's theorem from quantum mechanics. This theorem claims that the rules of classical mechanics are effective even for quantum mechanics if the classical physical quantities are replaced by quantum mechanical expectation values. For sufficiently accurate calculations of the expectation values, the range of the spatial integral must be sufficiently larger than the optical wavelength. However, for the case discussed in the present section, this range is not large enough because of the sub-wavelength size of the fiber probe tip, which drastically decreases the accuracy of calculation.

where the excitation energies of a probe, a sample, and a macroscopic material system are $\hbar\Omega_0(p)$, $\hbar\Omega_0(s)$, and $\hbar\Omega$, respectively, and the effective mass of the exciton–polariton is denoted by m_{P}. Equation (8.15) is transformed using these effective masses, and the magnitude of the effective interaction V_{eff} is written as a function of the inter-central distance r between the sample (atom in the present case) and probe, which is expressed by the sum of Yukawa functions Y as [9.13]

$$V_{\mathrm{eff}}(r) = P_+Y(\Delta_{\mathrm{P}+}r) - P_-Y(\Delta_{\mathrm{P}-}r) + S_+Y(\Delta_{\mathrm{S}+}r) - S_-Y(\Delta_{\mathrm{S}-}r)\,, \tag{9.8a}$$

$$Y(\kappa r) \equiv \frac{\mathrm{e}^{-\kappa r}}{r}\,. \tag{9.8b}$$

Since we consider atom manipulation using the optical near field, only the terms representing the optical near fields, i.e., the first and third terms of (8.15), have been used to derive these equations. Constants $P_\pm$ and $S_\pm$ are independent of r and are expressed as

$$P_\pm \propto \pm\frac{\hbar[\Omega \pm \Omega_0(p)][\Omega \pm \Omega_0(p)]^2}{[\Omega_0(p) \pm W_+\Omega][\Omega_0(p) \pm W_-\Omega]}\,, \tag{9.9a}$$

$$S_\pm \propto \pm\frac{\hbar[\Omega \pm \Omega_0(s)][\Omega \pm \Omega_0(s)]^2}{[\Omega_0(s) \pm W_+\Omega][\Omega_0(s) \pm W_-\Omega]}\,, \tag{9.9b}$$

where the constants $W_\pm$ are independent of energies. Since $|\Delta_{\mathrm{P}+}| > |\Delta_{\mathrm{P}-}|$ and $|\Delta_{\mathrm{S}+}| > |\Delta_{\mathrm{S}-}|$, the interaction ranges of the terms $Y(\Delta_{\mathrm{P}+}r)$ and $Y(\Delta_{\mathrm{S}+}r)$ are shorter than those of $Y(\Delta_{\mathrm{P}-}r)$ and $Y(\Delta_{\mathrm{S}-}r)$, respectively.

For a quantitative discussion of the magnitude of the interaction $V_{\mathrm{eff}}(r)$ and its r dependence, the transition from the $5S_{1/2}$ to $5P_{3/2}$ levels in $^{85}\mathrm{Rb}$ is taken as an example. The energy difference between these levels is 1.59 eV, which corresponds to $\hbar\Omega_0(s)$. Assuming the use of semiconductors, the values of $\hbar\Omega_0(p)$ and $\hbar\Omega$ are varied in the ranges $1.0\,\mathrm{eV} \leq \hbar\Omega_0(p) \leq 1.2\,\mathrm{eV}$ and $1.0\,\mathrm{eV} \leq \hbar\Omega \leq 1.8\,\mathrm{eV}$ in order to calculate the value of $V_{\mathrm{eff}}(r)$. Figures 9.8a–c show the results of this calculation. Curves A and B in Fig. 9.8a represent the terms $P_+Y(\Delta_{\mathrm{P}+}r)$ and $S_+Y(\Delta_{\mathrm{S}+}r)$ of (9.8a), respectively. Curve C is their sum, i.e., $V_{\mathrm{eff}}(r)$. They are derived for $\hbar\Omega_0(p) = 1.2\,\mathrm{eV}$, $\hbar\Omega = 1.0\,\mathrm{eV}$, and for red detuning $\delta = \Omega - \Omega_0(s) < 0$. In this case, it is found from (9.9a) and (9.9b) that P_+ and S_+ are negative, and thus, $V_{\mathrm{eff}}(r)$ is also negative. This means that $V_{\mathrm{eff}}(r)$ forms a potential attracting an atom toward the probe tip.

On the other hand, curves A, B, and C in Figs. 9.8b and c represent the terms $S_+Y(\Delta_{\mathrm{S}+}r) - S_-Y(\Delta_{\mathrm{S}-}r)$, $P_+Y(\Delta_{\mathrm{P}+}r) - P_-Y(\Delta_{\mathrm{P}-}r)$, and their sum $V_{\mathrm{eff}}(r)$, respectively. Figure 9.8b shows these curves for the case of $\hbar\Omega_0(p) = 1.0\,\mathrm{eV}$, $\hbar\Omega = 1.8\,\mathrm{eV}$, and blue detuning $\delta > 0$. Figure 9.8c is for $\hbar\Omega_0(p) = 1.2\,\mathrm{eV}$, $\hbar\Omega = 1.8\,\mathrm{eV}$, and blue detuning $\delta > 0$. The two figures show that the value of $V_{\mathrm{eff}}(r)$ depends on the value of each term of (9.8a). The negative $S_+Y(\Delta_{\mathrm{S}+}r) - S_-Y(\Delta_{\mathrm{S}-}r)$ term (see curve A in both

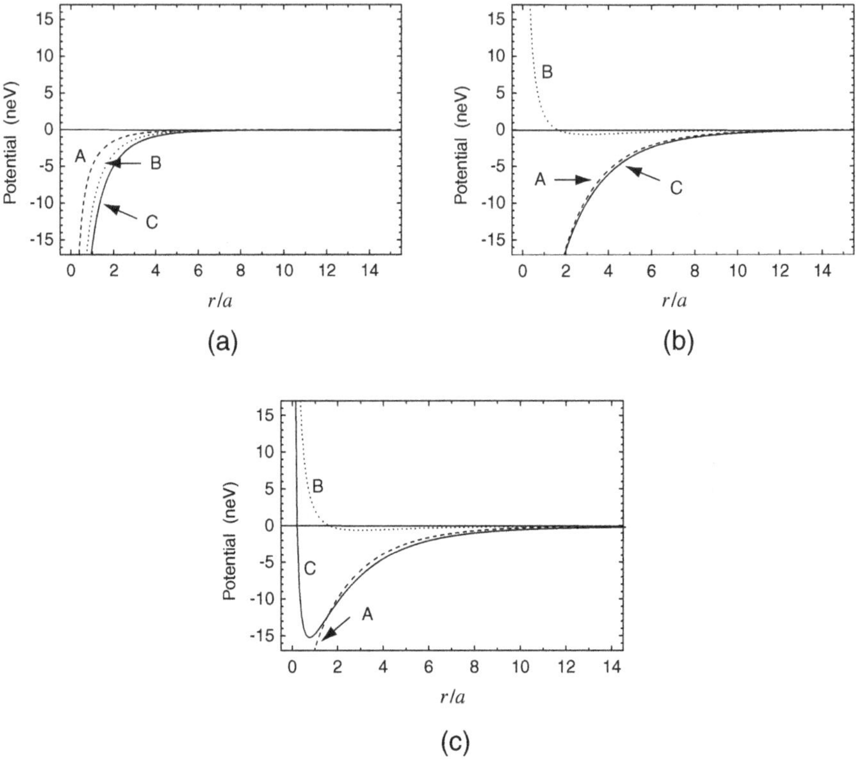

Fig. 9.8. Magnitude of the potential due to an optical near field. (**a**) Calculated results for $\hbar\Omega_0(p) = 1.0\,\text{eV}$, $\hbar\Omega = 1.0\,\text{eV}$, and $\delta < 0$. Curves A and B represent the terms $P_+Y(\Delta_{P+}r)$ and $S_+Y(\Delta_{S+}r)$ of (9.8a), respectively. Curve C is $V_{\text{eff}}(r)$. (**b**) Calculated results for $\hbar\Omega_0(p) = 1.0\,\text{eV}$, $\hbar\Omega = 1.8\,\text{eV}$, and $\delta > 0$. Curves A, B, and C represent the terms $S_+Y(\Delta_{S+}r) - S_-Y(\Delta_{S-}r)$, $P_+Y(\Delta_{P+}r) - P_-Y(\Delta_{P-}r)$, and their sum $V_{\text{eff}}(r)$ in (9.8a), respectively. (**c**) Calculated results for $\hbar\Omega_0(p) = 1.2\,\text{eV}$, $\hbar\Omega = 1.8\,\text{eV}$, and $\delta > 0$. Curves A, B, and C represent the terms $S_+Y(\Delta_{S+}r) - S_-Y(\Delta_{S-}r)$, $P_+Y(\Delta_{P+}r) - P_-Y(\Delta_{P-}r)$, and their sum $V_{\text{eff}}(r)$ in (9.8a), respectively

figures) shows that the atom is attracted to the probe. On the other hand, the positive $P_+Y(\Delta_{P+}r) - P_-Y(\Delta_{P-}r)$ term (curve B) shows that the atom is repelled from the probe. Since $|S_+Y(\Delta_{S+}r) - S_-Y(\Delta_{S-}r)| > |P_+Y(\Delta_{P+}r) - P_-Y(\Delta_{P-}r)|$ in the case of Fig. 9.8b, the magnitude of the total interaction $V_{\text{eff}}(r)$ (curve C) corresponds to an attractive potential. In the case of Fig. 9.8c, since $|S_+Y(\Delta_{S+}r) - S_-Y(\Delta_{S-}r)| > |P_+Y(\Delta_{P+}r) - P_-Y(\Delta_{P-}r)|$ and $|S_+Y(\Delta_{S+}r) - S_-Y(\Delta_{S-}r)| < |P_+Y(\Delta_{P+}r) - P_-Y(\Delta_{P-}r)|$ for larger and smaller r, respectively, the value of $V_{\text{eff}}(r)$ takes its minimum at a certain r, i.e., a potential well is formed.

The above discussions suggest that the potential profile, i.e., the dependence of $V_{\text{eff}}(r)$ on r, can be tailored by selecting the materials and sizes of the probe tip in order to control the thermal motion of an atom. This means that the potential profile can be tailored to trap a single atom in the near-field region of the probe tip. According to such tailoring capability, atom trapping and also deflection can be discussed by approximating the probe tip as a sphere with radius a for simplicity. By the same concept as was used to derive (8.16), the magnitude of the potential at the position $\boldsymbol{r}_{\mathrm{A}}$ of an atom due to the probe sphere P at $\boldsymbol{r}_{\mathrm{P}}$ is derived. It is expressed as the sum of Yukawa functions [9.13]:

$$\begin{aligned} V(r) &= \frac{1}{4\pi a^3/3} \int V_{\mathrm{eff}}\left[|\boldsymbol{r}_{\mathrm{A}} - (\boldsymbol{r}' + \boldsymbol{r}_{\mathrm{P}})|\right] \mathrm{d}^3 r' \\ &= Y_0 \sum_{G=P}^{S} \sum_{g=\mathrm{P}}^{\mathrm{S}} \sum_{j=-}^{+} \frac{jG_j}{\Delta_{gj}^3} \left[(1 + a\Delta_{gj})\mathrm{e}^{-\Delta_{gj}a} - (1 - a\Delta_{gj}a)\right] Y(\Delta_{gj} r) \\ &\equiv \sum_{G=P}^{S} \sum_{g=\mathrm{P}}^{\mathrm{S}} \sum_{j=-}^{+} jZ_{0j} G_j Y(\Delta_{gj} r) , \end{aligned} \tag{9.10}$$

where Y_0 and Z_{0j} are constants.

Deflection of an Atom

Slowly moving atoms, which are prepared by the method of laser cooling, are used for atom deflection experiments. When the equivalent temperature of the atomic thermal motion is cooled down to 10 mK, the velocity of the atoms is as low as 1 m/s. Such atoms are scattered by the potential described in (9.10). As an observable physical quantity, the differential scattering cross-section $\sigma(\theta)$ is derived using quantum mechanics. It represents the fractional flux of atoms scattered in the deflection angle θ. For this derivation, the momentum transfer K is defined as

$$K = 2\frac{Mv}{\hbar} \sin\frac{\theta}{2} , \tag{9.11}$$

in terms of the mass M and velocity v of an incident atom. By applying the Born approximation to the wave function of an atomic beam, $\sigma(\theta)$ is given by [9.14]

$$\sigma(\theta) = \left| -\frac{1}{4\pi} \left(\frac{8\pi M}{K\hbar^2}\right) \int_a^\infty r\mathrm{d}r V(r) \sin(Kr) \right|^2 . \tag{9.12}$$

Substituting (9.10) into this equation, one derives [9.15]

$$\begin{aligned} \sigma(\theta) &= \left| \left(\frac{2M}{K\hbar^2}\right) \sum_{G=P}^{S} \sum_{g=\mathrm{P}}^{\mathrm{S}} \sum_{j=-}^{+} jZ_{0j} G_j \int_a^\infty \mathrm{d}r \mathrm{e}^{-\Delta_{gj} r} \sin(Kr) \right|^2 \\ &= \left| \left(\frac{2M}{K\hbar^2}\right) \sum_{G=P}^{S} \sum_{g=\mathrm{P}}^{\mathrm{S}} \sum_{j=-}^{+} jZ_{0j} G_j \frac{K\cos(Ka) + \Delta_{gj}\sin(Ka)}{K^2 + \Delta_{gj}^2} \right|^2 . \end{aligned} \tag{9.13}$$

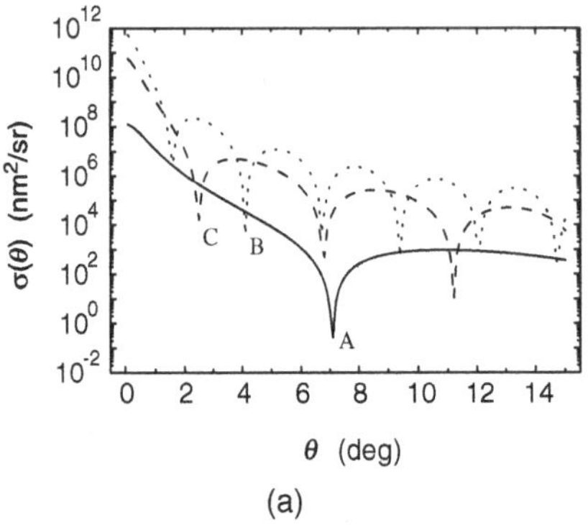

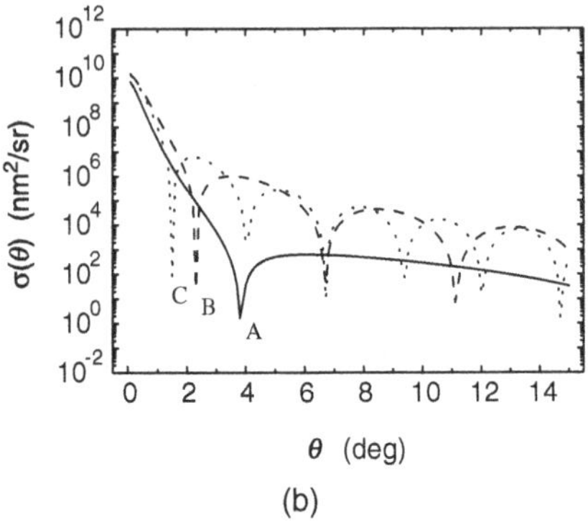

Fig. 9.9. Calculated differential cross-section for ^{85}Rb atom scattering. (**a**) and (**b**) represent the calculated results for scattering by the potential of curve C in Figs. 9.8a and b, respectively. The curves A, B, and C in (**a**) and (**b**) are the results for the probe sphere P with radius $a = 10$, 30, and 50 nm, respectively

Figures 9.9a and b show the value of $\sigma(\theta)$ for ^{85}Rb atoms with an incident velocity of 1 m/s, injected into the potentials expressed by the curves C in Figs. 9.8a and b, respectively. The curves A, B, and C in the figures represent the results for sphere P of radius $a = 10$, 30, and 50 nm, respectively. In both figures, three curves have periodic structures due to the finite size of the probe, whose period is inversely proportional to a. The value of the deflection angle θ at the first local minimum of the differential cross-section also increases with decreasing a. The difference in the profiles of the curves in Figs. 9.8a and b depends on whether all the terms in $V(r)$ of (9.10) take the same sign or not. That is, each term has the same sign for Fig. 9.9a when summing over G and j ($G = P, S, j = +, -$) in (9.13), which results in a larger deflection angle θ for the first local minimum. For Fig. 9.9b, this angle is smaller because the signs of the terms are opposite to each other.

To estimate typical values of the deflection angle θ and displacement x (see Fig. 9.7a for their definitions), the incident atomic beam flux N and minimum detectable flux N_{d} are assumed to be $10^{10}\,\mathrm{cm}^{-2}\mathrm{s}^{-1}$ and $10^{3}\,\mathrm{s}^{-1}$, respectively. To detect the signal, the condition $N\sigma(\theta) \geq N_{\mathrm{d}}$ must be satisfied. In the case $a = 10$ nm, the curve A in Fig. 9.9b shows that this condition is satisfied if $\theta \leq 1°$. If $z = 10$ cm in Fig. 9.7a, this deflection angle leads to $x = 1.8$ mm, which is sufficiently large to be measured by a conventional atom detector.

Trapping an Atom

Figure 9.7b shows how to trap an atom using a potential well, which is plotted as the curve C in Fig. 9.8c. In this case, the atom and the probe tip form a system which is equivalent to a diatomic molecule. The atom receives the attractive and repulsive forces simultaneously in this system. By balancing

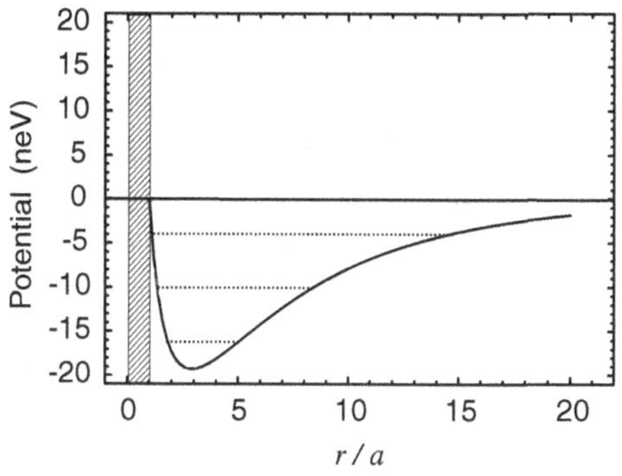

Fig. 9.10. Profile of the potential well for a ^{85}Rb atom due to an optical near field generated by the probe sphere P with radius $a = 10\,\text{nm}$. The center of sphere P corresponds to $r = 0$, which means that the *hatched area* is inside the probe

the two forces, the atom can exist at the bottom of the potential well and is thereby trapped. This trapping mechanism is similar to the balance between the attractive dipole force and repulsive cavity QED force discussed using conventional propagating light.

Figure 9.10 shows the profile of a potential well for a ^{85}Rb atom generated by the probe sphere ($a = 10\,\text{nm}$). Here the transition from the $5S_{1/2}$ level to the $5P_{3/2}$ level is assumed, whose transition energy is $1.59\,\text{eV} = \hbar\Omega_0(s)$. The excitation energies of the sphere P and the macroscopic material system were taken as $\hbar\Omega_0(p) = 1.51\,\text{eV}$ and $\hbar\Omega = 2.0\,\text{eV}$, respectively. This figure shows that the potential has a minimum at the position $2a$ from the probe tip surface. Three horizontal lines in the potential well represent the vibrational energy levels with quantum numbers $n = 0, 1$, and 2 of the equivalent harmonic oscillator calculated by approximating the bottom of the potential well as a parabola. The equivalent temperatures of the thermal motion of these vibration energies are several tens of μK, which means that if a ^{85}Rb atom is coded down to these equivalent temperatures, it can be trapped by the optical near field generated at the probe tip.

Future Development

Various developments are expected from the atom manipulation discussed above. For example, experimental results on the atom deflection provide interesting information about the atom–optical-near-field interaction. The profile of the optical near field on the fiber probe tip has usually been measured via destructive methods using a fiber probe (see Fig. 2.6 of Chap. 2). However, using atoms instead of the fiber probe as the secondary probe, non-destructive measurement of the optical near field is possible. The scheme for this non-destructive measurement enables one to examine the atom–optical-near-field interaction as a scattering problem or an inverse scattering problem. At the same time, the validity of the Yukawa potential model of Sect. 8.2 can be assessed through this measurement, which is important for understanding the local electromagnetic interaction in a nanometric space.

The cavity QED effect in the near-field region is also an important factor, but remains an open question. Single-molecule spectroscopy and atom manipulation discussed in Sects. 9.1 and 9.2, respectively, try to study the cavity QED effect experimentally using a fiber probe. The atom deflection scheme described above can be used as a powerful tool for investigating attractive forces due to the cavity QED effect, comparing it with the dipole force.

Precise control of atomic thermal motion gives rise to fruitful studies in related areas. As one can bring atoms to any position on a substrate using a hollow optical fiber, it becomes feasible to deposit a single atom, which is also useful in surface science and the semiconductor industry. For example, it is possible to grow an atomic-scale silicon crystal with optical-near-field devices using light tuned to the guide wavelength of 252 nm. In addition, near-field optical devices for atom manipulation are excellent tools for isotope separation, as seen in Sect. 3.3.5. This technique is a powerful way to obtain high accuracy in controlling the number and purity of neutral atoms in the isotope separation process.

Moreover, the optical near field generated at a fiber probe tip can be used as a quantum tweezer for an atom, which can be applied to novel material fabrication. These advanced techniques in atom manipulation using an optical near field will thus open new areas of scientific research, with quantum mechanical and optical objectives, and industrial applications including nanofabrication.

9.3 Nanophotonic Switching

Miniaturization of optical devices beyond the diffraction limit of light is not possible as long as propagating light is used, which was pointed out in Sect. 1.3. In order to realize this miniaturization to ensure progress in optical technology, the development of nanophotonics is indispensable [9.16]. Figure 9.11 shows the concept and structure of a nanophotonic integrated circuit. It is composed of a variety of nanoparticles which work as a nanolight emitter, a nanophotonic switch, and so on. Plasmon waveguides with nanometric width are also implemented for input/output terminals [9.17]. The nanophotonic switch is one of the essential ingredients in a nanophotonic integrated circuit. However, in the case of conventional optical switches, miniaturization of devices is limited by diffraction because it uses propagating light as a signal carrier. However, if one uses the optical near field as a signal carrier, miniaturization of optical devices can be achieved beyond the diffraction limit. Here it should be noted that several considerations, e.g., unidirectional signal transmission and reduced cross-talk between input and output signals,

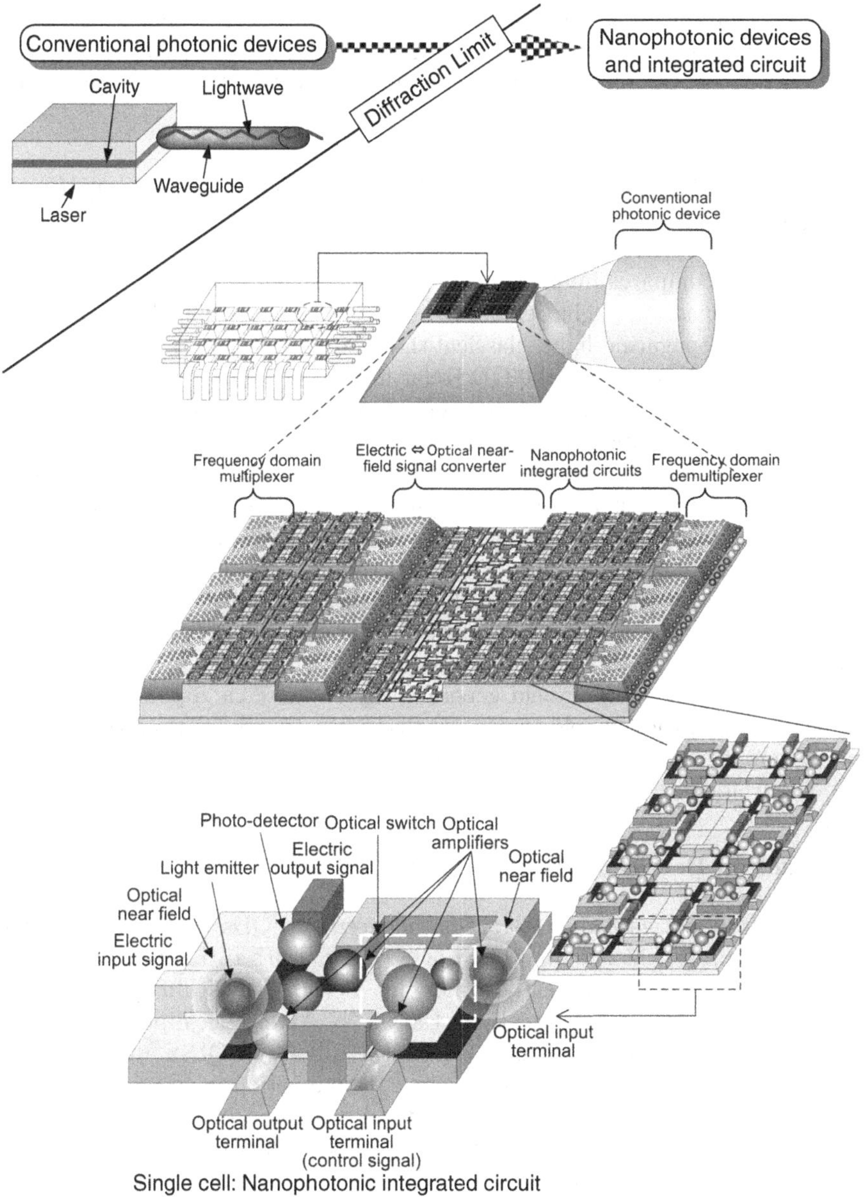

Fig. 9.11. Concept and structure of a nanophotonic integrated circuit

must be guaranteed when designing nanophotonic devices due to the non-propagating nature of the optical near field. Noting these considerations, this section reviews the function of a nanophotonic switch, based on the theory of Chap. 8 [9.16, 9.18–9.20].

9.3.1 Interaction and Energy Transfer Between Quantum Dots via Optical Near Field

This section compares interactions between quantum dots (QDs) via a conventional propagating light field and via an optical near field.[6] When a QD is excited by propagating light, classical theory explains that an electric dipole is induced at the center of the QD (refer to Chaps. 4 and 5) and the electric field generated from this electric dipole is detected in the far-field region. Quantum theory explains that an electron in the QD is excited from the ground state to an excited state due to the interaction between the electric dipole and electric field of the propagating light, which is called an electric dipole transition. A photon is then emitted by the transition from the excited state to the ground state, and detected in the far-field region.

In the following discussion, to distinguish the interaction via an optical near field from the one via a propagating far field, it is assumed that two anti-parallel electric dipoles are induced in a QD, as shown in Fig. 9.12a.[7] In this situation, the electric field generated by one electric dipole is cancelled by the other in the far-field region, and thus the transition from the excited state cannot take place. Then, the transition and excited state are said to be dipole-forbidden.

Interaction Between Quantum Dots

In order to investigate the interaction between two QDs via an optical near field, we assume that one quantum dot S (QD-S) with two anti-parallel electric dipoles is located near the other dot P (QD-P) within its near field region, as explained schematically in Fig. 9.12b. An electric dipole is induced in QD-P by the optical near field due to the dipoles in QD-S and generates another optical near field. One can detect the transition in QD-S by measuring the optical near field using the method described in Sect. 2.2, which is the dipole-forbidden transition in the case mediated by propagating light. Its quantum mechanical description is that QD-P changes the spatial distribution of the electric field around QD-S and induces an additional interaction whose magnitude is proportional to the gradient of the electric field.[8] Due

[6] Note that, although quantum dots are used as an example of a nanometric material in this discussion, the quantum dot may be replaced by any relevant nanometric material for more general discussions.

[7] This pair of electric dipoles is called an electric quadrupole, as explained in Sect. D.1.

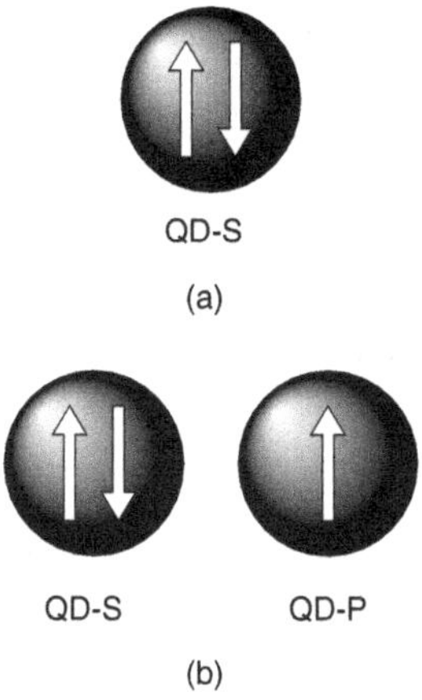

Fig. 9.12. Arrangements of electric dipoles (*white arrows*) induced in quantum dots. (**a**) Two anti-parallel electric dipoles induced in a quantum dot S. (**b**) When a quantum dot P is located near S, an electric dipole is induced in P, which is anti-parallel to one of the electric dipoles in S

to this change, the transition from the dipole-forbidden state in QD-S to the ground state can take place. This is a unique interaction attributed to the optical near field.

In order to estimate the magnitude of the effective interaction between QD-P and QD-S on the basis of the theory of Chap. 8, the two QDs are regarded as the subsystem (N), in which they correspond to the probe and the sample, respectively. Here we consider a case that the transition from the ground state to the dipole-forbidden excited state in QD-S. It is the inverse process of the one discussed above. Assuming that QD-P interacts with the macroscopic subsystem (M) via the incident light, the situation corresponds to those represented by Figs. 8.3c and d because QD-P is locally excited by the exciton–polariton. Therefore, as is the case when deriving (9.8a) and (9.8b), the magnitude of the effective interaction[9] $V_{\text{eff}}(r)$ between the two QDs is expressed by two Yukawa functions $Y(\Delta_{\text{S}+}r)$ and $Y(\Delta_{\text{P}-}r)$:

$$V_{\text{eff}}(r) = S_{+}Y(\Delta_{\text{S}+}r) - P_{-}Y(\Delta_{\text{P}-}r)\,, \tag{9.14}$$

where r, $\Delta_{\text{S}+}$, $\Delta_{\text{P}-}$, S_{+}, and P_{-} are the distance between the centers of the two QDs, and constants given in (9.7), (9.9a), and (9.9b), respectively. This expression includes the dipole-forbidden interaction described above.

Excitation Energy Transfer

This section describes intrinsic phenomena due to the effective interaction $V_{\text{eff}}(r)$, i.e., excitation energy transfer and nutation. First, we study the case

[8] In fact it corresponds to the interaction due to the 2^l-pole moment ($l = 2$) described in Sect. D.1.

[9] The discussion in Chap. 8 explains that this effective interaction originates in the exchange of a virtual exciton–polariton between the two QDs.

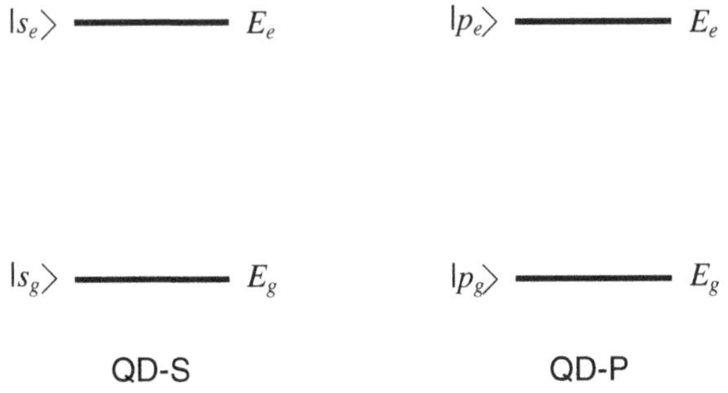

Fig. 9.13. A system composed of two quantum dots P and S with resonant energy levels

where the two QDs are isolated from each other. State vectors of the ground state and excited state of the electron in QD-P are expressed as $|p_{\mathrm{g}}\rangle$ and $|p_{\mathrm{e}}\rangle$, respectively, as shown in Fig. 9.13.[10] Ground and excited states in QD-S are expressed as $|s_{\mathrm{g}}\rangle$ and $|s_{\mathrm{e}}\rangle$. These state vectors are assumed to form an orthonormal set. Energy eigenvalues for $|p_{\mathrm{g}}\rangle$ and $|s_{\mathrm{g}}\rangle$ are expressed as E_{g}, while those for $|p_{\mathrm{e}}\rangle$ and $|s_{\mathrm{e}}\rangle$ are expressed as E_{e}. The states $|p_{\mathrm{g}}\rangle$ and $|s_{\mathrm{g}}\rangle$ are said to be resonant with each other or resonant levels because of their equal energy eigenvalues. The states $|p_{\mathrm{e}}\rangle$ and $|s_{\mathrm{e}}\rangle$ are also resonant with each other.

When two isolated QDs are placed close enough to induce the effective interaction $V_{\mathrm{eff}}(r)$, the energy eigenvalues and state vectors are modified. They are expressed as

$$E_{\pm} = E_{\mathrm{g}} + E_{\mathrm{e}} \pm V_{\mathrm{eff}}(r) \,, \tag{9.15a}$$

$$|\varphi_{\pm}\rangle = \frac{1}{\sqrt{2}}(|p_{\mathrm{e}}\rangle|s_{\mathrm{g}}\rangle \pm |p_{\mathrm{g}}\rangle|s_{\mathrm{e}}\rangle) \,, \tag{9.15b}$$

respectively. It follows that the states $|\varphi_{\pm}\rangle$ of (9.15b) also form an orthonormal set. The state vector for the system $|\Psi(t)\rangle$ at time t is expressed by the superposition of $|\varphi_{\pm}\rangle$ as

$$|\Psi(t)\rangle = \frac{1}{\sqrt{2}}\left[\exp\left(-\frac{\mathrm{i}E_{+}t}{\hbar}\right)|\varphi_{+}\rangle + \exp\left(-\frac{\mathrm{i}E_{-}t}{\hbar}\right)|\varphi_{-}\rangle\right] \,. \tag{9.16}$$

It is assumed that $|\Psi(0)\rangle$ is orthonormalized and that the electron in QD-P occupies the excited state at $t = 0$, while the electron in QD-S occupies the ground state, i.e.,

$$|\Psi(0)\rangle = |p_{\mathrm{e}}\rangle|s_{\mathrm{g}}\rangle \,. \tag{9.17}$$

Substituting this and (9.15) into (9.16), one obtains

[10] Although the behavior of an electron is discussed here, the electron may be replaced by an exciton, as will be done in the next section.

$$|\Psi(t)\rangle = \exp\left(-\frac{\mathrm{i}\bar{E}t}{\hbar}\right)\left[\cos\left(\frac{V_{\mathrm{eff}}(r)t}{\hbar}\right)|p_{\mathrm{e}}\rangle|s_{\mathrm{g}}\rangle - \mathrm{i}\sin\left(\frac{V_{\mathrm{eff}}(r)t}{\hbar}\right)|p_{\mathrm{g}}\rangle|s_{\mathrm{e}}\rangle\right] , \tag{9.18a}$$

$$\bar{E} = \frac{E_+ + E_-}{2} = E_{\mathrm{g}} + E_{\mathrm{e}} . \tag{9.18b}$$

From these equations, the probability that the electrons in QD-P and QD-S occupy the excited and ground states, respectively, is expressed as

$$\rho_{p_{\mathrm{e}}s_{\mathrm{g}}}(t) = |\langle s_{\mathrm{g}}|\langle p_{\mathrm{e}}|\Psi(t)\rangle|^2 = \cos^2\left(\frac{V_{\mathrm{eff}}(r)t}{\hbar}\right) . \tag{9.19}$$

Similarly, the probability that the electrons in QD-P and QD-S occupy the ground and excited states, respectively, is written as

$$\rho_{p_{\mathrm{g}}s_{\mathrm{e}}}(t) = |\langle s_{\mathrm{e}}|\langle p_{\mathrm{g}}|\Psi(t)\rangle|^2 = \sin^2\left(\frac{V_{\mathrm{eff}}(r)t}{\hbar}\right) . \tag{9.20}$$

These equations show that the probability varies periodically with period $t = \pi\hbar/V_{\mathrm{eff}}(r)$. In other words, the excitation energy of this system is periodically transferred between the two QDs. This is called nutation.[11]

Guaranteeing Unidirectional Signal Transmission

Unidirectional transmission of the input signal from QD-P to QD-S can be disturbed by the nutation discussed above.[12] To guarantee unidirectivity, a fast mechanism is required to dissipate part of the excitation energy after it is transferred to QD-S. This mechanism can be realized by using the fact that the electron in a QD has discrete energy levels due to the quantum confinement effect.

In the case of a semiconductor QD, E_{g} and E_{e} represent the electron energies in the valence and conduction bands, respectively, where E_{e} is one of the discrete energies. By the interaction between the electron and phonon, the energy in the conduction band can be dissipated to a state with lower energy eigenvalue than E_{e}. Since the relaxation time for this dissipation is as short as about 1 ps, unidirectional signal transmission can be guaranteed by appropriately adjusting this relaxation time and the time constant of the interaction between the two QDs via the optical near field.

9.3.2 Principle and Operation of a Nanophotonic Switch

A nanophotonic switch is composed of three cubic quantum dots (QD1, 2, and 3). As explained schematically in Fig. 9.14a, the sides of these cubes

[11] The term 'excitation energy transfer' is equivalent to transfer of 'the electron energy difference in the excited and ground states between QD-S and QD-P'.

[12] Indeed, the term 'unidirectional' means that the excitation energy is transferred only from QD-P to QD-S.

are $a/2$, $a/\sqrt{2}$, and a, respectively, i.e., their ratio is $1 : \sqrt{2} : 2$. Further, the distances between adjacent QDs are maintained shorter than a in order to induce a sufficiently strong interaction via the optical near field, called the optical-near-field interaction. The behavior of electrons was discussed in the previous section, and we now investigate that of excitons. Because the distance between the electron and hole is shorter than the QD size in the case of CuCl to be discussed in the next section, it is more convenient to treat the electron–hole pair as an exciton. The energy levels of the exciton are shown in Fig. 9.14b. The energy eigenvalues in the QD with size a are expressed as

$$E(a; n_x, n_y, n_z) = \frac{\pi^2 \hbar^2}{2Ma^2}(n_x^2 + n_y^2 + n_z^2) \, , \quad n_x, n_y, n_z = 1, 2, \dots \, , \tag{9.21}$$

which represent the discrete energy levels described in the previous section.[13] Since the size ratio of the three QDs is $1 : \sqrt{2} : 2$, there are several energy levels which are resonant between adjacent QDs, e.g., the energy level $E(a/2, n_x = n_y = n_z = 1)$ of QD1 is resonant with $E(a/\sqrt{2}, n_x = 2, n_y = n_z = 1)$ of QD2 and $E(a, n_x = n_y = n_z = 2)$ of QD3. From now onward, QD1, 2, and 3 will be referred to as the input QD (QD-I), output QD (QD-O), and control QD (QD-C), respectively.

An exciton can be transferred from level $E(a/2, n_x = n_y = n_z = 1)$ of QD-I to level $E(a/\sqrt{2}, n_x = 2, n_y = n_z = 1)$ of QD-O by near-field optical interaction, as described in Sect. 9.3.1. Similar transfer is also possible, e.g.,

$$\begin{aligned}
&E(a/2, n_x = n_y = n_z = 1) \ \text{of QD-I} \\
&\qquad\qquad \longrightarrow E(a, n_x = n_y = n_z = 2) \ \text{of QD-C} \, , \\
&E(a/\sqrt{2}, n_x = 2, n_y = n_z = 1) \ \text{of QD-O} \\
&\qquad\qquad \longrightarrow E(a, n_x = n_y = n_z = 2) \ \text{of QD-C} \, , \\
&E(a/\sqrt{2}, n_x = n_y = n_z = 1) \ \text{of QD-O} \\
&\qquad\qquad \longrightarrow E(a, n_x = 2, n_y = n_z = 1) \ \text{of QD-C} \, ,
\end{aligned}$$

as indicated by solid arrows in Fig. 9.14b. The signal is transferred in the form of energy transfer.

However, unidirectional signal transmission is not guaranteed because an exciton can be transferred in the opposite direction due to nutation between the resonant levels. In order to avoid this back-transfer, a relaxation process is used which originates in the interaction between the exciton and phonon. Broken arrows in Fig. 9.14b represent this process. Since its time constant is rather short, the exciton energies in level $E(a/\sqrt{2}, n_x = 2, n_y = n_z = 1)$ of QD-O and level $E(a, n_x = n_y = n_z = 2)$ of QD-C are immediately dissipated to the phonon system, and the exciton transits to a lower energy level. As a result of this transition, back-transfers of the exciton from the QD-O to

[13] The bulk exciton energy E_B is subtracted from this energy.

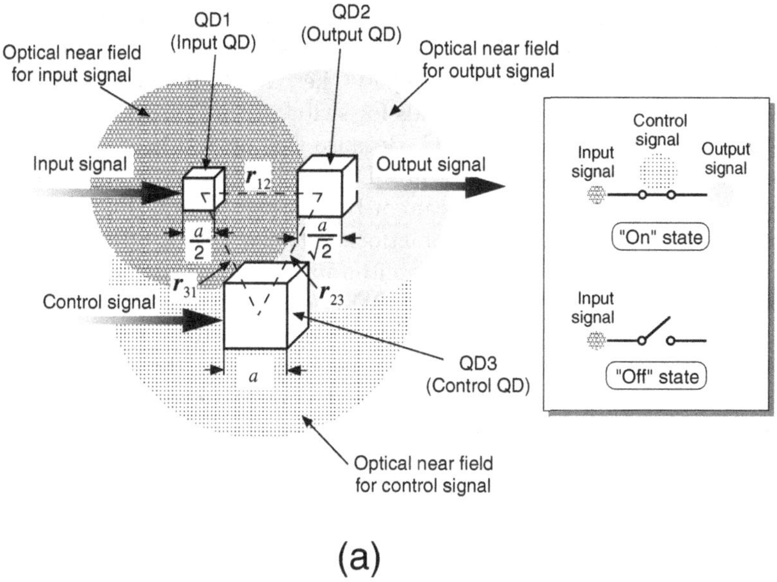

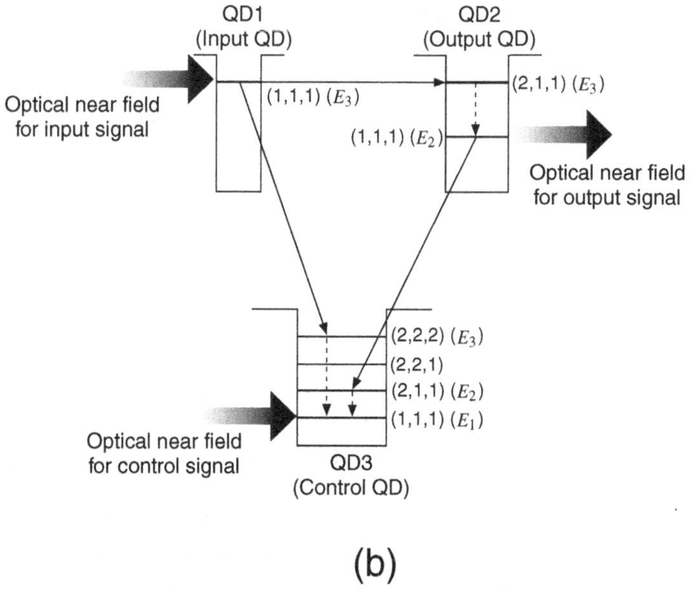

Fig. 9.14. Structure of a nanophotonic switch. (**a**) Three cubic quantum dots used for the switch and their arrangements. (**b**) Energy levels of the exciton and their quantum numbers (n_x, n_y, n_z) in each quantum dot. *Solid* and *broken arrows* represent the transition by optical-near-field interaction and relaxation due to the interaction with a phonon, respectively

QD-I, and from QD-C to QD-I (or QD-O) are avoided. Further, cross-talk between input and output signals is also avoided due to the difference in the exciton energies between QD-I and QD-O, i.e., due to frequency conversion.

Control and modulation of signals for switching are carried out by applying the optical near field to QD-C. Because this optical near field, i.e., the control signal, excites an exciton in QD-C to its $E(a, n_x = n_y = n_z = 1)$ level, excitons in QD-I or QD-O cannot be transferred to QD-C. Due to this state filling, optical-near-field interactions between QD-C, QD-I, and QD-O are prohibited, and as a result the exciton in QD-I can only transfer to QD-O. After this transfer, the exciton in QD-O generates the optical near field, which is detected and transferred to the external photonic system. This situation corresponds to the 'on' state of the switch, i.e., the switch is closed. On the other hand, the situation in which the optical near field is not applied to QD-C corresponds to the 'off' state, i.e., the switch is open, because the exciton energy is transferred to the lowest level of QD-C and the optical near field is not obtained from the QD-O.

In order to investigate the switching described above, the energies $E(a/2, n_x = n_y = n_z = 1)$ of QD-I, $E(a/\sqrt{2}, n_x = 2, n_y = n_z = 1)$ of QD-O, and $E(a/\sqrt{2}, n_x = 2, n_y = n_z = 1)$ of QD-O are expressed as E_3, E_2, and E_3, respectively, for simplicity. It follows from (9.21) that the energies $E(a, n_x = n_y = n_z = 1)$, $E(a, n_x = 2, n_y = n_z = 1)$, and $E(a, n_x = n_y = n_z = 2)$ of QD-C can also be expressed as E_1, E_2, and E_3, respectively. We use a differential equation describing the temporal behavior of the probability $P_n^j(t)$, representing the fact that the energy level E_n in QD-j is occupied by an exciton. It is expressed as

$$\begin{cases} \dfrac{\mathrm{d}P_n^j(t)}{\mathrm{d}t} = -\left[\dfrac{1}{\tau_n} + (U_n^{ij})^2\right] P_n^j(t) + (U_n^{jk})^2 P_n^k(t)\,, \\ U_n^{ij} = \dfrac{Y_1}{\hbar} \dfrac{\exp[-m_{\mathrm{eff}}(n) r_{ij}]}{m_{\mathrm{eff}}(n) r_{ij}}\,, \end{cases} \tag{9.22}$$

where τ_n represents the relaxation time of the transition of the exciton from level E_n of one QD to the lower level E_{n-1} due to the interaction with phonons. We denote Y_1 as a constant and $(U_n^{ij})^2$ as the energy transfer rate between the resonant levels E_n of the adjacent QDs due to the optical-near-field interaction. The rate $(U_n^{ij})^2$ is expressed as a Yukawa function using the effective mass $m_{\mathrm{eff}}(n)$ of the exciton–polariton for the transition between the resonant levels E_n of QD-i and QD-j. The distance between the centers of the two QDs is represented by r_{ij}. The effective mass $m_{\mathrm{eff}}(n)$ determines the effective range in which the optical-near-field interaction can excert.

To investigate the temporal behavior of the probability $P_n^j(t)$, CuCl QDs are a good example because their emission efficiency is very high. Assuming that $a = 10\,\mathrm{nm}$, $r_{ij} = 10\,\mathrm{nm}$, and $\tau_n = 1\,\mathrm{ps}$ for CuCl QDs, the temporal behavior of the probability $P_2^2(t)$ of occupying the level E_2 of QD-O is obtained using the initial condition $P_3^1(0) = 1$ (i.e., the exciton occupies level E_3 of

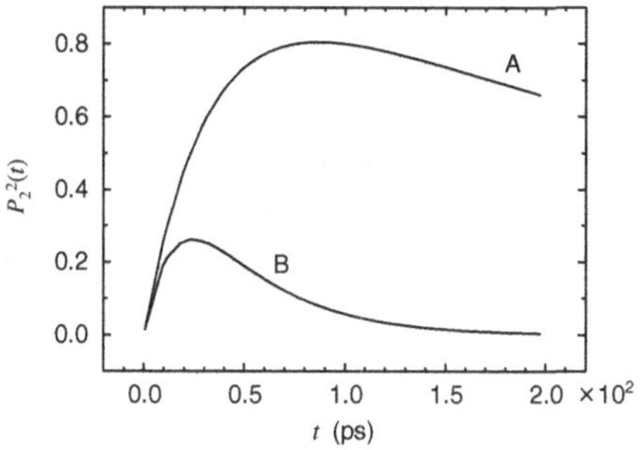

Fig. 9.15. Temporal behavior of the probability $P_2^2(t)$. Curves A and B represent the calculated results in the respective cases when the optical near field is or is not applied to the control quantum dot (QD-C)

QD-I at $t = 0$). Figure 9.15 shows the calculated result. Curve A in this figure shows that the value of $P_2^2(t)$ saturates within about 60 ps after applying the optical-near-field signal to QD-C. Curve B shows that $P_2^2(t)$ approaches 0 within 60–70 ps after turning the signal off. A typical switching time is estimated to be several 10–100 ps. Similarly to the case of CuCl, switching characteristics are also expected for ZnO, GaAs, and GaN. The readers who are interested in the details are referred to [9.21, 9.22], where the dynamic properties are discussed by using a quantum master equation.

9.3.3 Experiments to Confirm Nanophotonic Switching

In order to fabricate the nanophotonic switch described in the previous section, the QDs have to be deposited on a substrate, controlling their sizes and positions. Photochemical vapor deposition by an optical near field is a promising method for this purpose, as was reviewed in Sect. 3.3.3. Experiments have been carried out using QDs fabricated by a conventional technique as a preliminary study for confirming principles of the switching. Such experiments should:

- identify the sizes and positions of the QDs,
- confirm the optical-near-field energy transfer between the QDs,[14]
- confirm the switching operation.

These experiments have been carried out for CuCl QDs fabricated in a NaCl host crystal by photoluminescence spectroscopy at low temperature, as described in Sect. 3.3.2.

[14] Although the exciton energy is transferred in the case of CuCl QDs, other transfer mechanisms are also expected in different nanometric materials.

Experiment (1)

Figure 9.16a shows photoluminescence spectral profiles at 18 K measured using a 325 nm wavelength He–Cd laser to excite the CuCl QDs. Curve A is a photoluminescence spectrum measured by means of conventional spectroscopy using propagating light for excitation. It shows inhomogeneous broadening due to the distributed sizes of the QDs. On the other hand, curve

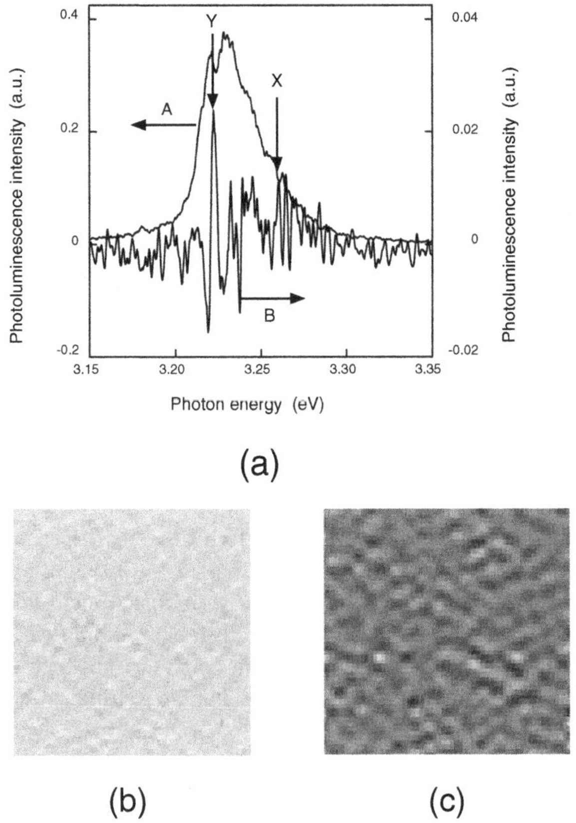

Fig. 9.16. Photoluminescence spectra emitted from CuCl quantum dots fabricated in a NaCl host crystal. (**a**) Photoluminescence spectral profile. Curve A shows a photoluminescence spectrum of the CuCl quantum dots measured by means of conventional spectroscopy using propagating light for excitation. Curve B is a photoluminescence spectrum measured using near-field-optical spectroscopy, which was described in Sect. 3.3.2. (**b**) Spatial distribution of the photoluminescence intensity emitted from a QD of size 3.0 nm. The photoluminescence energy is indicated as the peak X in (**a**). (**c**) Spatial distribution of the photoluminescence intensity for a QD of size 4.9 nm. The photoluminescence energy is indicated as the peak Y in (**a**). Image sizes in (**b**) and (**c**) are both 600 nm $\times$ 600 nm

B is a photoluminescence spectrum measured using near-field-optical spectroscopy, which shows fine structures due to the discrete energy levels of individual CuCl QDs.

Figures 9.16b and c show the spatial distributions of the two kinds of CuCl QDs in the size ratio $1 : \sqrt{2}$. Figure 9.16b shows the spatial distribution of the photoluminescence intensity emitted from a QD with size 3.0 nm, whose photoluminescence energy is indicated as the peak X in Fig. 9.16a. Similarly, Fig. 9.16c shows the spatial distribution of the photoluminescence intensity for a QD with size 4.9 nm, whose photoluminescence energy is indicated as the peak Y in Fig. 9.16a. Noting that the size ratio of these QDs is $1 : \sqrt{2}$, these results indicate that we can identify the sizes and positions of QDs.

Experiment (2)

Figures 9.17a and b show the spatial distribution of the photoluminescence intensities measured from two kinds of CuCl QDs with sizes 3.9 nm and 5.6 nm, respectively. Note that the size ratio is also $1 : \sqrt{2}$, as was the case in Figs. 9.16b and c. The areas surrounded by broken curves in Fig. 9.17a are dark (i.e., photoluminescence is not observed), while the corresponding areas in Fig. 9.17b are bright. This anti-correlation feature means that excitons are transferred from the CuCl QDs with size 3.9 nm to those with size 5.6 nm, and as a result only the photoluminescence from the 5.6 nm CuCl QDs is detected. This feature corresponds to the transfer of an exciton from level E_3 of QD-I to level E_3 of QD-O, due to the optical-near-field interaction. Then QD-O emits light after relaxation to level E_2, due to the interaction with phonons, as shown in Fig. 9.14b.

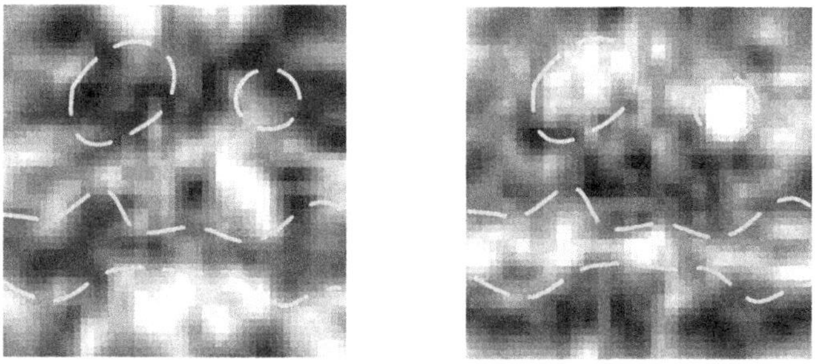

(a) (b)

Fig. 9.17. Spatial distribution of the photoluminescence intensities emitted from CuCl quantum dots fabricated in a NaCl host crystal. Image sizes are both 200 nm × 200 nm. (**a**) and (**b**) represent the measured results for CuCl quantum dots with sizes 3.9 nm and 5.6 nm, respectively

Experiment (3)

Time-resolved pump–probe spectroscopy is carried out using pulsed lasers. The switching time was measured as 62 ps in the case of CuCl QDs [9.23]. This agrees with the theoretical results obtained in the previous section.

Problems

Problem 9.1

Derive (9.2) from (9.1).

Problem 9.2

Derive (9.5) from (9.1).

A Basic Formulae of Electromagnetism

We will present the basic formulae of electromagnetism used throughout this book. Readers are referred to the textbooks on electromagnetism for the details concerning derivations.

A.1 Maxwell's Equations and Related Formulae

A.1.1 Static Electric and Magnetic Fields

An electric field $\boldsymbol{E}$ is a gradient of a scalar potential ϕ, i.e.,

$$\boldsymbol{E} = -\nabla\phi \,. \tag{A.1}$$

The operator ∇ called nabla is given by

$$\nabla = \boldsymbol{i}\frac{\partial}{\partial x} + \boldsymbol{j}\frac{\partial}{\partial y} + \boldsymbol{k}\frac{\partial}{\partial z} \,,$$

in the case of a Cartesian coordinate system. Unit vectors along x, y, and z axes are represented by $\boldsymbol{i}$, $\boldsymbol{j}$, and $\boldsymbol{k}$, respectively. The potential ϕ generated by true electric charges with volume density ρ in vacuum is given by Coulomb's law, viz.,

$$\phi(\boldsymbol{r}) = \frac{1}{4\pi\varepsilon_0}\int \frac{\rho(\boldsymbol{r}')}{|\boldsymbol{r}-\boldsymbol{r}'|}\mathrm{d}^3r' \,, \tag{A.2}$$

where ε_0 is the dielectric constant in vacuum ($= 8.85418782 \times 10^{-12}\,\mathrm{F/m}$). Gauss's law is

$$\nabla\cdot\boldsymbol{E} = \frac{\rho}{\varepsilon_0} \,, \tag{A.3}$$

which represents the fact that the electric lines of force originate from a positive charge and terminate on a negative charge. The quantity $\nabla\cdot\boldsymbol{E}$ on the left-hand side of (A.3) is the scalar product of the operator ∇ and the electric field $\boldsymbol{E}$, which is given by

$$\nabla\cdot\boldsymbol{E} = \frac{\partial E_x}{\partial x} + \frac{\partial E_y}{\partial y} + \frac{\partial E_z}{\partial z} \,,$$

where $\boldsymbol{E}$ is expressed as (E_x, E_y, E_z) in the Cartesian coordinate system. In an arbitrary medium, this equation is modified to $\nabla\cdot\boldsymbol{D} = \rho$, where $\boldsymbol{D}$ is the electric flux density described in (A.15).

A differential equation for ϕ is derived from (A.1) and (A.3):

$$\Delta\phi = -\frac{\rho}{\varepsilon_0} \,. \tag{A.4}$$

This is Poisson's equation. The operator Δ is $\nabla\cdot\nabla$ or ∇^2, which is called the Laplacian. In a Cartesian coordinate system, $\Delta\phi$ is expressed as

$$\Delta\phi = \left(\frac{\partial^2}{\partial x^2} + \frac{\partial^2}{\partial y^2} + \frac{\partial^2}{\partial z^2}\right)\phi \,.$$

In the absence of charges ($\rho = 0$), (A.4) is called Laplace's equation.

The law of conservation of electric charge is

$$\frac{\partial\rho}{\partial t} + \nabla\cdot\boldsymbol{j} = 0 \,, \tag{A.5}$$

where $\boldsymbol{j}$ is the electric current density. This equation, which is called the continuity equation, represents conservation and continuity of electric charges.

In order to present the basic formulas for the magnetic field, a vector potential $\boldsymbol{A}$ is defined by

$$\boldsymbol{A} \equiv \frac{\mu_0}{4\pi}\int \frac{\boldsymbol{j}(\boldsymbol{r}')}{|\boldsymbol{r} - \boldsymbol{r}'|}\mathrm{d}^3 r' \,, \tag{A.6}$$

where $\boldsymbol{j}$ acts as the source of the magnetic field. It is easily understood that the magnetic flux density $\boldsymbol{B}$ can be written in the form

$$\boldsymbol{B} = \nabla \times \boldsymbol{A} \,, \tag{A.7}$$

since substituting (A.6) into (A.7) leads to the Biot–Savart law. The quantity $\nabla \times \boldsymbol{A}$ is the vector product of the operator ∇ and vector $\boldsymbol{A}$. It is defined by

$$\nabla \times \boldsymbol{A} = \boldsymbol{i}\left(\frac{\partial A_z}{\partial y} - \frac{\partial A_y}{\partial z}\right) + \boldsymbol{j}\left(\frac{\partial A_x}{\partial z} - \frac{\partial A_z}{\partial x}\right) + \boldsymbol{k}\left(\frac{\partial A_y}{\partial x} - \frac{\partial A_x}{\partial y}\right)$$

in the Cartesian coordinate system, where $\boldsymbol{A}$ is expressed as (A_x, A_y, A_z).

The forms of (A.6) and (A.7) for the magnetic field are analogous to those of (A.2) and (A.1) for the electric field, respectively. Further, (A.7) is valid even when $\boldsymbol{A}$ is replaced by the sum of $\boldsymbol{A}$ and an arbitrary scalar function ψ because $\nabla \times \nabla\psi = 0$. This is called gauge invariance. Given this freedom, one can select an appropriate scalar function ψ for the static magnetic field so that the relation

$$\nabla\cdot\boldsymbol{A} = 0 \tag{A.8}$$

is valid.

Ampere's law is expressed as

$$\nabla \times \boldsymbol{B} = \mu_0 \boldsymbol{j} , \tag{A.9}$$

for the magnetic field generated by $\boldsymbol{j}$, where μ_0 is the magnetic permeability in vacuum ($= 1.25663706 \times 10^{-6}$ H/m). By substituting (A.7) into this equation and using the mathematical formula $\nabla \times (\nabla \times \boldsymbol{A}) = -\Delta \boldsymbol{A} + \nabla(\nabla \cdot \boldsymbol{A})$, one derives

$$\Delta \boldsymbol{A} - \nabla(\nabla \cdot \boldsymbol{A}) = -\mu_0 \boldsymbol{j} . \tag{A.10}$$

Using (A.8), this transforms to

$$\Delta \boldsymbol{A} = -\mu_0 \boldsymbol{j} , \tag{A.11}$$

which has the same form as Poisson's equation (A.4) for the scalar potential.

A.1.2 Dynamic Electric and Magnetic Fields

Faraday's law of electromagnetic induction is

$$\nabla \times \boldsymbol{E} = -\frac{\partial \boldsymbol{B}}{\partial t} , \tag{A.12}$$

which says that variation of the magnetic field generates an electric field. Ampere's law tells us that an electric current generates a magnetic field $\boldsymbol{H}$:

$$\nabla \times \boldsymbol{H} = \boldsymbol{j} + \frac{\partial \boldsymbol{D}}{\partial t} , \tag{A.13}$$

which is a more general form than (A.9). The second term on the right-hand side is called the displacement current and $\boldsymbol{D}$ is an electric flux density. The spatial profile of the magnetic flux is illustrated schematically as a closed loop because magnetic monopoles do not exist. It is expressed as

$$\nabla \cdot \boldsymbol{B} = 0 . \tag{A.14}$$

The set of four equations (A.12)–(A.14) together with (A.3) are called Maxwell's equations. Further, electric and magnetic properties of a medium are given by

$$\boldsymbol{D} = \varepsilon \boldsymbol{E} , \tag{A.15a}$$

$$\boldsymbol{B} = \mu \boldsymbol{H} , \tag{A.15b}$$

where ε and μ are the dielectric constant and magnetic permeability of the medium, respectively. These equations are called the medium equations. Further, (A.15a) and (A.15b) can be written

$$\boldsymbol{D} = \varepsilon_0 \boldsymbol{E} + \boldsymbol{P} , \tag{A.16a}$$

$$\boldsymbol{B} = \mu_0 (\boldsymbol{H} + \boldsymbol{M}) , \tag{A.16b}$$

by introducing the polarization $\boldsymbol{P}$ and magnetic charge $\boldsymbol{M}$, respectively.

The electric field generated by $\boldsymbol{P}$ is

$$\nabla \times \nabla \times \boldsymbol{E} + \varepsilon_0 \mu_0 \frac{\partial^2 \boldsymbol{E}}{\partial t^2} = -\mu_0 \frac{\partial^2 \boldsymbol{P}}{\partial t^2} , \tag{A.17}$$

where (A.12), (A.13), and (A.16a) are used under the condition $\boldsymbol{B} = \mu_0 \boldsymbol{H}$ and $\boldsymbol{j} = 0$. Equation (A.1) must be modified for the case of dynamic electric and magnetic fields because the relation $\nabla \times \boldsymbol{E} = 0$ derived from (A.1) is inconsistent with (A.12). However, the relation $\boldsymbol{E} + \partial \boldsymbol{A}/\partial t = -\nabla \phi$ is valid because substitution of (A.7) into (A.12) leads to $\nabla \times (\boldsymbol{E} + \partial \boldsymbol{A}/\partial t) = 0$. (This relation is derived by noting an identical equation $\nabla \times \nabla \phi = 0$ for an arbitrary scalar function ϕ.) Thus, (A.1) can be modified to

$$\boldsymbol{E} = -\frac{\partial \boldsymbol{A}}{\partial t} - \nabla \phi . \tag{A.18}$$

In the case of a medium in which $\boldsymbol{j} = 0$, $\boldsymbol{P} = 0$, and $\boldsymbol{M} \neq 0$, note that

$$\nabla \times \nabla \times \boldsymbol{A} + \varepsilon_0 \mu_0 \frac{\partial^2 \boldsymbol{A}}{\partial t^2} = \mu_0 \nabla \times \boldsymbol{M} . \tag{A.19}$$

This is derived by applying $\nabla \times$ to (A.16b) and using (A.7), (A.13), (A.16a), and (A.18), where $\nabla \phi$ is assumed to be zero. This equation is used in Chap. 7.

A.1.3 Electromagnetic Fields Generated by an Electric Dipole

The temporal and spatial variation of the dynamic electric and magnetic fields are assumed to take the form $\exp(-\mathrm{i}\omega t + \mathrm{i}kr)$, where ω and k are angular frequency and wave number, respectively. They are expressed as $\omega = 2\pi\nu$ and $k = 2\pi/\lambda$ in terms of the frequency ν and wavelength λ. Further, the phase velocity v of the electromagnetic fields is expressed as $v = \nu\lambda = \omega/k$. Noting that the relations $\boldsymbol{j} = 0$, $\boldsymbol{D} = \varepsilon_0 \boldsymbol{E}$, and $B = \mu_0 \boldsymbol{H}$ are valid in vacuum, we transform (A.13) to

$$\boldsymbol{E} = \mathrm{i}\frac{c^2}{\omega} \nabla \times \boldsymbol{B} = \mathrm{i}\frac{c}{k} \nabla \times \nabla \times \boldsymbol{A} , \tag{A.20}$$

where the relations $\partial/\partial t = -\mathrm{i}\omega$ and $c = 1/\sqrt{\mu_0 \varepsilon_0}$ (the phase velocity of the electromagnetic fields in vacuum, viz., 2.99792458×10^8 m/s) have been used. Further, (A.7) was used to transform from the middle to the right-hand side.

For electromagnetic fields with the form $\exp(-\mathrm{i}\omega t + \mathrm{i}kr)$, (A.6) becomes

$$\boldsymbol{A} = \frac{\mu_0}{4\pi} \int \frac{\boldsymbol{j}(\boldsymbol{r}')\mathrm{e}^{\mathrm{i}k|\boldsymbol{r}-\boldsymbol{r}'|}}{|\boldsymbol{r} - \boldsymbol{r}'|} \mathrm{d}^3 r' , \tag{A.21}$$

which can be transformed to

$$\boldsymbol{A} = \frac{\mu_0}{4\pi}\frac{\mathrm{e}^{ikr}}{r}\int \boldsymbol{j}(\boldsymbol{r}')\mathrm{d}^3r' . \tag{A.22}$$

Since the relation

$$\int \boldsymbol{j}(\boldsymbol{r}')\mathrm{d}^3r' = -\int \boldsymbol{r}'(\nabla_{\boldsymbol{r}'}\cdot\boldsymbol{j})\mathrm{d}^3r' \tag{A.23}$$

is valid, where $\nabla_{\boldsymbol{r}'}$ is the differential operator with respect to $\boldsymbol{r}'$, substitution of (A.5) into this equation gives

$$\int \boldsymbol{j}(\boldsymbol{r}')\mathrm{d}^3r' = -\mathrm{i}\omega\int \boldsymbol{r}'\rho(\boldsymbol{r}')\mathrm{d}^3r' . \tag{A.24}$$

Here $\partial/\partial t = -\mathrm{i}\omega$ has been used. Defining the electric dipole moment $\boldsymbol{p}$ by

$$\boldsymbol{p} \equiv \int \boldsymbol{r}'\rho(\boldsymbol{r}')\mathrm{d}^3r' , \tag{A.25}$$

(A.22)–(A.24) lead to

$$\boldsymbol{A} = -\frac{\mathrm{i}\omega\mu_0}{4\pi}\boldsymbol{p}\frac{\mathrm{e}^{ikr}}{r} . \tag{A.26}$$

Equation (A.25) describes the case of a continuous spatial distribution of electric charge with volume density $\rho(\boldsymbol{r}')$. In the case of a pair of positive and negative point charges $+q$ and $-q$ separated by a distance $\boldsymbol{d}$, the electric dipole moment is given by $\boldsymbol{p} = q\boldsymbol{d}$. (The reader is recommended to refer to the solution to Problem 4.1 of Chap. 4 and Fig. Q4.1 for the derivation). Substituting (A.26) into (A.7), one derives

$$\boldsymbol{B} = \frac{\omega\mu_0 k}{4\pi}(\boldsymbol{n}\times\boldsymbol{p})\frac{\mathrm{e}^{ikr}}{r}\left(1-\frac{1}{ikr}\right) , \tag{A.27}$$

where $\boldsymbol{n}$ is a unit vector along $\boldsymbol{r}$, i.e., $\boldsymbol{n} = \boldsymbol{r}/|\boldsymbol{r}| = (x/r, y/r, z/r)$. Substituting this into (A.20), one arrives at

$$\boldsymbol{E} = \frac{1}{4\pi\varepsilon_0}\left\{k^2(\boldsymbol{n}\times\boldsymbol{p})\times\boldsymbol{n}\left(\frac{1}{r}\right) + [3\boldsymbol{n}(\boldsymbol{n}\cdot\boldsymbol{p})-\boldsymbol{p}]\left(-\frac{ik}{r^2}+\frac{1}{r^3}\right)\right\}\mathrm{e}^{ikr} . \tag{A.28}$$

The first, second, and third terms on the right-hand side of this equation are proportional to $(kr)^{-1}$, $(kr)^{-2}$, and $(kr)^{-3}$, respectively. The first term represents the component that dominates in the far-field region, because its magnitude is the largest among the three terms when $kr \gg 1$. On the other hand, the third term represents the electric field component which dominates in the proximity of $\boldsymbol{p}$, because its magnitude is the largest among the three terms when $kr \ll 1$. In order to investigate the characteristics of the electric field represented by this equation, it is assumed that $\boldsymbol{p}$ is fixed at the origin in the Cartesian coordinate system (xyz) and is oriented along the y-axis, i.e.,

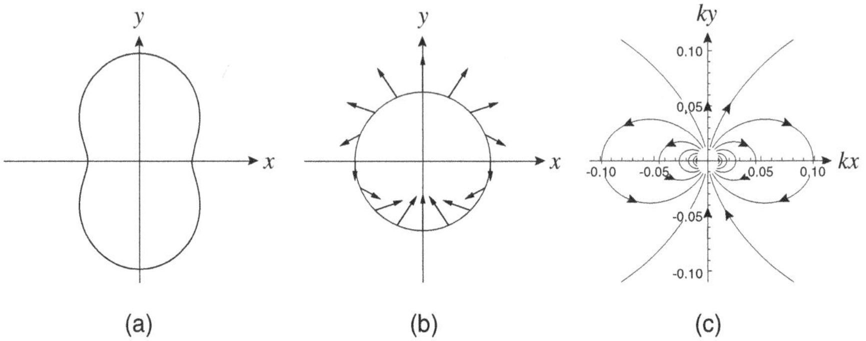

Fig. A.1. Electric field generated by an electric dipole $\boldsymbol{p}$ in its near-field region ($kr \ll 1$), where $\boldsymbol{p}$ is fixed at the origin and oriented along the y-axis. (**a**) Absolute value of the vector $3\boldsymbol{n}(\boldsymbol{n}\cdot\boldsymbol{p})-\boldsymbol{p}$. (**b**) Direction of the vector $3\boldsymbol{n}(\boldsymbol{n}\cdot\boldsymbol{p})-\boldsymbol{p}$. The *direction* of each *arrow* represents the direction of this vector at each position. The *length* of the *arrow* represents the absolute value of the vector, as illustrated in (**a**). (**c**) Electric lines of force. *Horizontal* and *vertical axes* represent k_x and k_y, respectively

$\boldsymbol{p}=(0,p,0)$. The characteristics of this equation are investigated in the xy-plane for simplicity, because the electric field is symmetric around the y-axis. Under this condition, $\boldsymbol{r}$ is assumed to lie in the xy-plane with orientation angle θ relative to the x-axis, i.e., $\boldsymbol{r}=r(\cos\theta,\sin\theta,0)$.

First, the vector in the third term of (A.28) is expressed as $3\boldsymbol{n}(\boldsymbol{n}\cdot\boldsymbol{p})-\boldsymbol{p}= p(3\cos\theta\sin\theta, 3\sin^2\theta-1,0)$. Its absolute value is $p\sqrt{3\sin^2\theta+1}$, which is illustrated in Fig. A.1a as a function of θ. The arrows in Fig. A.1b represent the direction of the vector $3\boldsymbol{n}(\boldsymbol{n}\cdot\boldsymbol{p})-\boldsymbol{p}$, which is illustrated on a circle by noting that its orientation angle is $\tan^{-1}[(1-3\cos 2\theta)/3\sin 2\theta]$. The curves in Fig. A.1c represent the electric lines of force illustrated using Figs. A.1a and b. The direction of the tangential line on each curve represents the direction of the electric field, whilst the density of curves represents the magnitude of the electric field.

Second, because the vector in the first term of (A.28) is expressed as $(\boldsymbol{n}\times\boldsymbol{p})\times\boldsymbol{n}=p(-\sin\theta\cos\theta,\cos^2\theta,0)$, Fig. A.2a represents its absolute value $p|\cos\theta|$ plotted as a function of θ. Arrows in Fig. A.2b represent the direction of this vector, which is illustrated on a circle by noting that its orientation angle is $\theta-\pi/2$.

Finally, Fig. A.3 shows the electric lines of force illustrated using all the terms on the right-hand side of (A.28). Curves near the origin are butterfly-shaped with two wings directed along the $\pm x$-axes, while curves in the outer region are closed loops. This means that the former and latter represent a non-propagating optical near field and propagating far-field light, respectively.

It should be noted that there exist electric lines of force on the y-axis of Fig. A.1c whose directions lie along the y-axis, whereas Fig. A.3 does not have

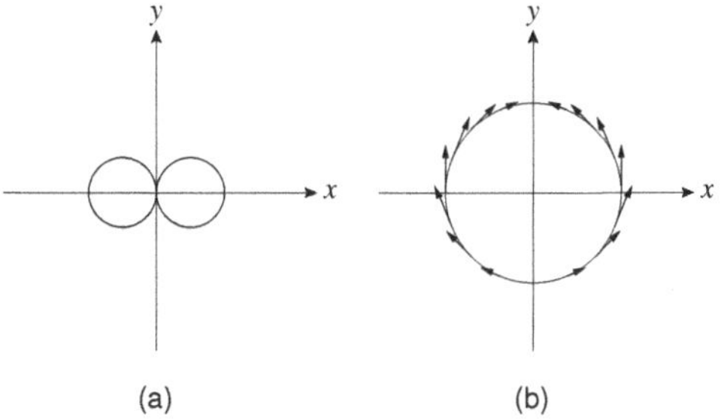

Fig. A.2. Electric field generated by an electric dipole $\boldsymbol{p}$ in its far-field region ($kr \gg 1$). The direction of $\boldsymbol{p}$ is the same as in Fig. A.1. (**a**) Absolute value of the vector $(\boldsymbol{n} \times \boldsymbol{p}) \times \boldsymbol{n}$. (**b**) Direction of the vector $(\boldsymbol{n} \times \boldsymbol{p}) \times \boldsymbol{n}$. The definition of the *arrows* is the same as in Fig. A.1b

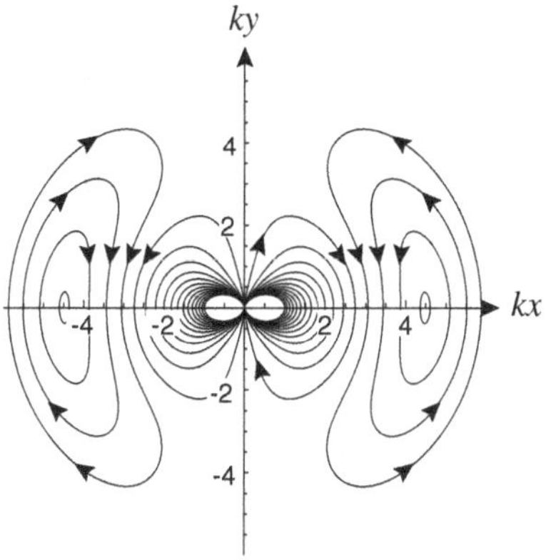

Fig. A.3. Electric lines of force of the total electric field generated by an electric dipole $\boldsymbol{p}$. *Horizontal* and *vertical axes* are represented by k_x and k_y, respectively. Note that the scale is different from the one in Fig. A.1c

these electric lines of force. This difference is due to the fact that the vector $(\boldsymbol{n} \times \boldsymbol{p}) \times \boldsymbol{n}$ in the first term of (A.28), representing the far-field electric field, is zero when $\boldsymbol{n}$ is parallel to $\boldsymbol{p}$. In other words, when the electric dipole $\boldsymbol{p}$ is observed from a distance along the y-axis, the positive and negative charges appear to cancelled each other and so the electric lines of force disappear.

Substituting $k = 0$ into (A.28), all the terms vanish except for $[3\boldsymbol{n}(\boldsymbol{n} \cdot \boldsymbol{p}) - \boldsymbol{p}](1/r^3)$. The non-vanishing term represents the static electric field, so the electric lines of force in Fig. A.1c are equivalent to those of the static electric field derived in Problem 4.1 of Chap. 4.

A.1.4 Power Radiated from an Electric Dipole

In order to obtain the time-averaged power P radiated from an oscillating electric dipole into the circumambient space, note that its value per unit solid angle is given by

$$\frac{\mathrm{d}P}{\mathrm{d}\Omega} = \frac{1}{2\mu_0}\mathrm{Re}(r^2 \boldsymbol{n} \cdot \boldsymbol{E} \times \boldsymbol{B}^*) , \tag{A.29}$$

where Re and $*$ represent the real part of the complex value in brackets and the complex conjugate, respectively. Since this power is detected in the far-field region of the electric dipole, the values of $\boldsymbol{E}$ and $\boldsymbol{B}$ are substituted into this equation under the approximation $kr \gg 1$. The approximate value of $\boldsymbol{B}$ given by (A.27) is

$$\boldsymbol{B} \approx \frac{\omega\mu_0 k}{4\pi}(\boldsymbol{n} \times \boldsymbol{p})\frac{\mathrm{e}^{\mathrm{i}kr}}{r} , \tag{A.30}$$

whilst that of $\boldsymbol{E}$ given by (A.28) is

$$\boldsymbol{E} \approx \frac{k^2}{4\pi\varepsilon_0}(\boldsymbol{n} \times \boldsymbol{p}) \times \boldsymbol{n}\frac{\mathrm{e}^{\mathrm{i}kr}}{r} . \tag{A.31}$$

Substituting (A.30) and (A.31) into (A.29), one derives

$$\frac{\mathrm{d}P}{\mathrm{d}\Omega} = \frac{k^3\omega}{32\pi^2\varepsilon_0}|(\boldsymbol{n} \times \boldsymbol{p}) \times \boldsymbol{n}|^2 . \tag{A.32}$$

The polarization of the radiated light is represented by the vector in $|\ |$, whose angular distribution is shown in Fig. A.2a. Substituting $|(\boldsymbol{n} \times \boldsymbol{p}) \times \boldsymbol{n}| = p|\cos\theta|$, (A.32) becomes

$$\frac{\mathrm{d}P}{\mathrm{d}\Omega} = \frac{k^3\omega}{32\pi^2\varepsilon_0}p^2\cos^2\theta . \tag{A.33}$$

Since $\mathrm{d}\Omega = \cos\theta \mathrm{d}\theta \mathrm{d}\phi$, (A.33) can be expressed as an integral, i.e.,

$$P = \int_0^{2\pi} \mathrm{d}\phi \int_{-\pi/2}^{\pi/2} \frac{k^3\omega}{32\pi^2\varepsilon_0}p^2\cos^3\theta \mathrm{d}\theta = \frac{p^2k^3\omega}{12\pi\varepsilon_0} . \tag{A.34}$$

Noting that the wave number k is inversely proportional to the velocity of light c (i.e., $k = \omega/c$), one obtains the result of integration as

$$P = \frac{p^2\omega^4}{12\pi\varepsilon_0 c^3} . \tag{A.35}$$

It shows that the radiated power is proportional to ω^4, which corresponds to the frequency dependence of the Rayleigh scattering intensity.

B Refractive Index of a Metal

Matter is generally composed of atoms, which possess positively charged nuclei attracting negatively charged electrons by a Coulomb force. Due to the attractive force of the nucleus, electrons hardly move around. However, in the case of atoms in a metal, the Coulomb force is screened due to the large number of electrons, so that electrons become free from the attractive force of the nucleus and move freely in the metal. This free movement of electrons in a metal makes it electrically conductive. Such electrons are called free electrons. This appendix derives the refractive index of a metal by analyzing the motion of free electrons due to the electric field of incident light.

The Drude model is the most popular theoretical model used for this analysis. It expresses the motion of an electron within the framework of classical mechanics. The equation of motion is given by

$$m\frac{\mathrm{d}^2 x}{\mathrm{d}t^2} = -\gamma\frac{\mathrm{d}x}{\mathrm{d}t} - eE_0\mathrm{e}^{-\mathrm{i}\omega t}\,, \tag{B.1}$$

where m is the mass, x the displacement, γ the damping constant due to scattering, and $-e$ the charge of the electron, respectively. The electric field of the incident light is expressed as $E_0\mathrm{e}^{-\mathrm{i}\omega t}$, where ω is the angular frequency.

Assuming an oscillatory displacement of the form $x = x_0\mathrm{e}^{-\mathrm{i}\omega t}$ and substituting it into (B.1), we have the amplitude x_0 as

$$x_0 = \frac{eE_0/m}{\omega^2 + \mathrm{i}(\gamma/m)\omega}\,. \tag{B.2}$$

Due to this displacement, an electric dipole moment $-ex_0$ is induced in each atom. Summing over all the electric dipole moments, the polarization $-ex_0N$ is generated in the metal, where N is the number of electrons per unit volume of the metal. This polarization is also expressed as $P = (\varepsilon - \varepsilon_0)E_0$, where ε and ε_0 are the dielectric constants of the metal and the vacuum, respectively. Thus the dielectric constant ε of the metal is derived from (B.2) as

$$\varepsilon = \varepsilon_0 - \frac{e^2N/m}{\omega^2 + \mathrm{i}(\omega/\tau)}\,, \tag{B.3}$$

where τ is defined by m/γ. Since $\varepsilon/\varepsilon_0$ is equal to the square of the refractive index n, one obtains from (B.3) the relation

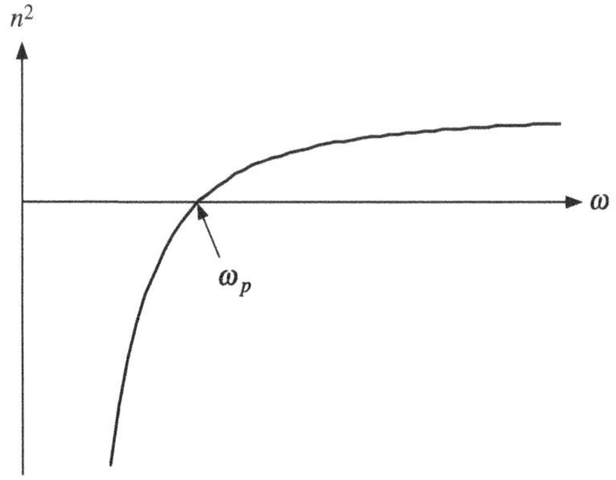

Fig. B.1. Relation between the angular frequency ω of incident light and the square of the refractive index n of a metal, where ω_{p} is the plasma angular frequency

$$n^2 = 1 - \frac{\omega_{\mathrm{p}}^2 \tau}{\omega(\omega\tau + \mathrm{i})} , \tag{B.4}$$

where ω_{p} is a constant called the plasma angular frequency given by

$$\omega_{\mathrm{p}} = \sqrt{\frac{e^2 N}{\varepsilon_0 m}} . \tag{B.5}$$

The quantity $\omega_{\mathrm{p}}/2\pi$ is called the plasma frequency.

The value of $\omega\tau$ is much larger than unity because the plasma frequency is higher than $2\pi \times 10^{15}$ Hz (i.e., in the visible or ultraviolet region) and the value of τ is about 1×10^{-14} s. Therefore the imaginary part of the right-hand side of (B.4) is negligible and one derives

$$n^2 = 1 - \left(\frac{\omega_{\mathrm{p}}}{\omega}\right)^2 . \tag{B.6}$$

Figure B.1 shows the relation between ω and n^2 of (B.6). Note that n^2 is negative if $\omega < \omega_{\mathrm{p}}$, which means that the refractive index n takes a purely imaginary value. This also means that the incident light is totally reflected on the surface of the metal. Since the approximation $\omega\tau \gg 1$ is valid for most metals, they have a high reflectivity and appear shiny.

C Exciton–Polariton

An exciton is the state in which an electron removed from a neutral atom stays in an orbit around the positively charged atom in a crystal, which is similar to the state of an electron in a free hydrogen atom. In the case of a semiconductor, a Coulomb force acts between an electron in the conduction band and a hole in the valence band. Therefore, an electron–hole pair behaves like a single particle. This pair is also called an exciton. When the distance between the electron and the hole (called the Bohr radius of the exciton) is shorter than the distance to adjacent atoms in the crystal, it is called a Frenkel exciton. When it is longer, the pair is called a Wannier exciton.

This appendix uses quantum theory to analyze the light–matter interaction based on the exciton concept. When light (photons) impinges on matter, it is absorbed and an exciton is created. This exciton is subsequently annihilated to create a photon. These processes are repeated in matter. Since photon creation and exciton annihilation take place repeatedly, a mixed state of photons and excitons must be described in order to analyze the light–matter interaction.

The repeated process means that a new stationary state is formed as a result of the interaction between a photon and an exciton. This state has an intrinsic energy dispersion relation. (A dispersion relation is a relation between the momentum and energy of a particle.) This state is regarded as a polarization field, which is an elementary excitation mode, and is called a polariton. Since we study a mixed state of the photon and exciton, it is called an exciton–polariton [C.1–C.5]. It is a coupled state of the electromagnetic field and the polarization field. Since the polarization with angular frequency ω_1 and the photon with angular frequency ω_2 are coupled, this state is similar to obtain a new oscillation with angular frequencies Ω_1 and Ω_2 by combining two harmonic oscillators.

The Hamiltonian of the exciton–polariton can be derived by representing the Hamiltonian of the light–electron interaction in the exciton picture. It is expressed as

$$\hat{H} = \sum_{\boldsymbol{k}} \hbar\omega_{\boldsymbol{k}} \hat{a}^{\dagger}_{\boldsymbol{k}} \hat{a}_{\boldsymbol{k}} + \sum_{\boldsymbol{k}} \hbar\varepsilon_{\boldsymbol{k}} \hat{b}^{\dagger}_{\boldsymbol{k}} \hat{b} + \sum_{\boldsymbol{k}} \hbar D(\hat{a}_{\boldsymbol{k}} + \hat{a}^{\dagger}_{-\boldsymbol{k}})(\hat{b}^{\dagger}_{\boldsymbol{k}} + \hat{b}_{-\boldsymbol{k}}) \,. \tag{C.1}$$

The first, second, and third terms represent the photon energy (energy eigenvalue $\hbar\omega_{\boldsymbol{k}}$ and angular frequency $\omega_{\boldsymbol{k}}$ corresponds to ω_1), the exciton energy

(energy eigenvalue $\hbar\varepsilon_{\boldsymbol{k}}$ and angular frequency $\varepsilon_{\boldsymbol{k}}$ corresponds to ω_2), and the interaction energy between the photon and exciton (magnitude of the interaction corresponds to $\hbar D$), respectively. The creation and annihilation operators of the photon are $\hat{a}_{\boldsymbol{k}}^{\dagger}$ and $\hat{a}_{\boldsymbol{k}}$, respectively. For the exciton, these operators are $\hat{b}_{\boldsymbol{k}}^{\dagger}$ and $\hat{b}_{\boldsymbol{k}}$, respectively, which are derived as follows. Operators $\hat{c}_{\boldsymbol{l},\mathrm{v}}$ and $\hat{c}_{\boldsymbol{l},\mathrm{c}}^{\dagger}$ annihilate an electron in the valence band and create an electron in the conduction band, respectively, at lattice site $\boldsymbol{l}$. Then the creation operator of an exciton at lattice site $\boldsymbol{l}$ is expressed as $\hat{b}_{\boldsymbol{l}}^{\dagger} = \hat{c}_{\boldsymbol{l},\mathrm{c}}^{\dagger}\hat{c}_{\boldsymbol{l},\mathrm{v}}$. Similarly, the annihilation operator is expressed as $\hat{b}_{\boldsymbol{l}} = \hat{c}_{\boldsymbol{l},\mathrm{v}}^{\dagger}\hat{c}_{\boldsymbol{l},\mathrm{c}}$. Taking all the lattice sites into account, $\hat{b}_{\boldsymbol{k}}^{\dagger}$ and $\hat{b}_{\boldsymbol{k}}$ are expressed as $(1/\sqrt{N})\sum_{\boldsymbol{l}}\exp(\mathrm{i}\boldsymbol{k}\cdot\boldsymbol{l})\hat{b}_{\boldsymbol{l}}^{\dagger}$, and $(1/\sqrt{N})\sum_{\boldsymbol{l}}\exp(-\mathrm{i}\boldsymbol{k}\cdot\boldsymbol{l})\hat{b}_{\boldsymbol{l}}$, respectively, where N represents the total number of lattice sites.

In order to obtain the eigenenergy of the exciton–polariton, the third term of (C.1) is expanded and the terms $\hat{a}_{\boldsymbol{k}}\hat{b}_{\boldsymbol{k}}^{\dagger}$ (annihilating the photon and creating the exciton) and $\hat{a}_{\boldsymbol{k}}^{\dagger}\hat{b}_{\boldsymbol{k}}$ (annihilating the exciton and creating the photon) are kept while the terms $\hat{a}_{\boldsymbol{k}}^{\dagger}\hat{b}_{-\boldsymbol{k}}^{\dagger}$ and $\hat{a}_{\boldsymbol{k}}\hat{b}_{-\boldsymbol{k}}$ (creating or annihilating the photon and the exciton simultaneously) are dropped. Under this approximation (a rotating wave approximation), $\hat{H}$ is expressed as

$$\hat{H} = \sum_{\boldsymbol{k}} \hat{H}_{\boldsymbol{k}} , \tag{C.2}$$

where

$$\hat{H}_{\boldsymbol{k}} = \hbar(\omega_{\boldsymbol{k}}\hat{a}_{\boldsymbol{k}}^{\dagger}\hat{a}_{\boldsymbol{k}} + \varepsilon_{\boldsymbol{k}}\hat{b}_{\boldsymbol{k}}^{\dagger}\hat{b}_{\boldsymbol{k}}) + \hbar D(\hat{b}_{\boldsymbol{k}}^{\dagger}\hat{a}_{\boldsymbol{k}} + \hat{a}_{\boldsymbol{k}}^{\dagger}\hat{b}_{\boldsymbol{k}}) . \tag{C.3}$$

In (C.3), the term $\hat{H}_{\boldsymbol{k}}$ is assumed to be diagonalized as

$$\begin{aligned}\hat{H}_{\boldsymbol{k}} &= \hbar(\Omega_{\boldsymbol{k},1}\hat{\xi}_1^{\dagger}\hat{\xi}_1 + \Omega_{\boldsymbol{k},2}\hat{\xi}_2^{\dagger}\hat{\xi}_2) = \hbar(\hat{b}_{\boldsymbol{k}}^{\dagger}, \hat{a}_{\boldsymbol{k}}^{\dagger})A\begin{pmatrix}\hat{b}_{\boldsymbol{k}}\\ \hat{a}_{\boldsymbol{k}}\end{pmatrix}\\ &= \hbar(a_{11}\hat{b}_{\boldsymbol{k}}^{\dagger}\hat{b}_{\boldsymbol{k}} + a_{12}\hat{b}_{\boldsymbol{k}}^{\dagger}\hat{a}_{\boldsymbol{k}} + a_{21}\hat{a}_{\boldsymbol{k}}^{\dagger}\hat{b}_{\boldsymbol{k}} + a_{22}\hat{a}_{\boldsymbol{k}}^{\dagger}\hat{a}_{\boldsymbol{k}}) ,\end{aligned} \tag{C.4}$$

using the creation (annihilation) operators $\hat{\xi}_1^{\dagger}$ and $\hat{\xi}_2^{\dagger}$ ($\hat{\xi}_1$ and $\hat{\xi}_2$) for two kinds of exciton–polaritons with angular frequencies Ω_1 and Ω_2, respectively. Comparing with (C.3), we can express the 2×2 matrix A in (C.4) as

$$A = \begin{pmatrix}a_{11} & a_{12}\\ a_{21} & a_{22}\end{pmatrix} = \begin{pmatrix}\varepsilon_{\boldsymbol{k}} & D\\ D & \omega_{\boldsymbol{k}}\end{pmatrix} . \tag{C.5}$$

After unitary transformation U of

$$\begin{pmatrix}\hat{b}_{\boldsymbol{k}}\\ \hat{a}_{\boldsymbol{k}}\end{pmatrix} = U\begin{pmatrix}\hat{\xi}_1\\ \hat{\xi}_2\end{pmatrix} = \begin{pmatrix}u_{11} & u_{12}\\ u_{21} & u_{22}\end{pmatrix}\begin{pmatrix}\hat{\xi}_1\\ \hat{\xi}_2\end{pmatrix} , \tag{C.6}$$

and substitution of (C.6) into the right-hand side of (C.4), one has

$$\hbar(\hat{b}_{\boldsymbol{k}}^{\dagger}, \hat{a}_{\boldsymbol{k}}^{\dagger})A\begin{pmatrix}\hat{b}_{\boldsymbol{k}}\\ \hat{a}_{\boldsymbol{k}}\end{pmatrix} = \hbar(\hat{\xi}_1^{\dagger}, \hat{\xi}_2^{\dagger})U^{\dagger}AU\begin{pmatrix}\hat{\xi}_1\\ \hat{\xi}_2\end{pmatrix} = \hbar(\hat{\xi}_1^{\dagger}, \hat{\xi}_2^{\dagger})U^{-1}AU\begin{pmatrix}\hat{\xi}_1\\ \hat{\xi}_2\end{pmatrix} . \tag{C.7}$$

(Note that a unitary transformation is characterized by the relation $U^\dagger U = 1$ or $U^\dagger = U^{-1}$.) Since the middle of (C.4) can be transformed to

$$\hbar(\Omega_{\boldsymbol{k},1}\hat{\xi}_1^\dagger\hat{\xi}_1 + \Omega_{\boldsymbol{k},2}\hat{\xi}_2^\dagger\hat{\xi}_2) = \hbar(\hat{\xi}_1^\dagger, \hat{\xi}_2^\dagger)\begin{pmatrix} \Omega_{\boldsymbol{k},1} & 0 \\ 0 & \Omega_{\boldsymbol{k},2} \end{pmatrix}\begin{pmatrix} \hat{\xi}_1 \\ \hat{\xi}_2 \end{pmatrix} , \tag{C.8}$$

we obtain

$$U^{-1}AU = \begin{pmatrix} \Omega_{\boldsymbol{k},1} & 0 \\ 0 & \Omega_{\boldsymbol{k},2} \end{pmatrix} \equiv \Lambda , \tag{C.9}$$

by equating (C.7) and (C.8). Then multiplying U from the left on both sides of this equation, it follows that $AU = U\Lambda$, which is expressed as

$$\begin{cases} (\varepsilon_{\boldsymbol{k}} - \Omega_{\boldsymbol{k},j})u_{1j} + Du_{2j} = 0 , \\ Du_{1j} + (\omega_{\boldsymbol{k}} - \Omega_{\boldsymbol{k},j})u_{2j} = 0 , \end{cases} \tag{C.10}$$

for $j = 1, 2$, or

$$\begin{pmatrix} \varepsilon_{\boldsymbol{k}} - \Omega_{\boldsymbol{k},1} & D \\ D & \omega_{\boldsymbol{k}} - \Omega_{\boldsymbol{k},j} \end{pmatrix}\begin{pmatrix} u_{1j} \\ u_{2j} \end{pmatrix} = 0 . \tag{C.11}$$

Since (u_{1j}, u_{2j}) on the left-hand side of (C.11) is a nonzero vector, the determinant of the 2×2 matrix on this side must be zero. From this requirement, one derives the secular equation

$$(\Omega_{\boldsymbol{k},j} - \varepsilon_{\boldsymbol{k}})(\Omega_{\boldsymbol{k},j} - \omega_{\boldsymbol{k}}) - D^2 = 0 . \tag{C.12}$$

The eigenenergies of the exciton–polariton are obtained by solving this equation to give

$$\hbar\Omega_{\boldsymbol{k},j} = \hbar\left[\frac{\varepsilon_{\boldsymbol{k}} + \omega_{\boldsymbol{k}}}{2} \pm \frac{\sqrt{(\varepsilon_{\boldsymbol{k}} - \omega_{\boldsymbol{k}})^2 + 4D^2}}{2}\right] . \tag{C.13}$$

Substituting the dispersion relation $\omega_{\boldsymbol{k}} = ck$ between the photon energy $\omega_{\boldsymbol{k}}$ and momentum $\hbar\boldsymbol{k}$ ($k = |\boldsymbol{k}|$) into (C.13), one derives the dispersion relation of an exciton–polariton, i.e., the relation between an eigenenergy and momentum of the exciton–polariton. This relation is represented schematically in Fig. C.1, in which the eigenvalue $\hbar\varepsilon_{\boldsymbol{k}}$ of the exciton energy is assumed to be constant $\hbar\Omega$ for simplicity.

It is found from (C.10) that the components u_{1j} and u_{2j} of the eigenvector satisfy the equation

$$u_{2j} = -\frac{\varepsilon_{\boldsymbol{k}} - \Omega_{\boldsymbol{k},j}}{D}u_{1j} . \tag{C.14}$$

Substituting this into $u_{1j}^2 + u_{2j}^2 = 1$ (which holds because U is unitary), one has

$$\left[1 + \left(\frac{\varepsilon_{\boldsymbol{k}} - \Omega_{\boldsymbol{k},j}}{D}\right)^2\right]u_{1j}^2 = 1 . \tag{C.15}$$

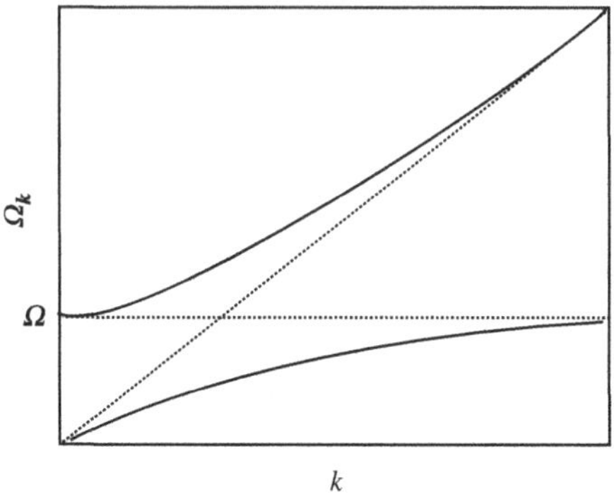

Fig. C.1. Dispersion relation of the exciton–polariton. The angular frequency Ω is proportional to the eigenenergy $\hbar\Omega$ of the exciton

Thus the eigenvector of the exciton–polariton is represented by

$$\begin{cases} u_{1j} = \left[1 + \left(\dfrac{\varepsilon_{\boldsymbol{k}} - \Omega_{\boldsymbol{k},j}}{D}\right)^2\right]^{-1/2}, \\ u_{2j} = -\left(\dfrac{\varepsilon_{\boldsymbol{k}} - \Omega_{\boldsymbol{k},j}}{D}\right)\left[1 + \left(\dfrac{\varepsilon_{\boldsymbol{k}} - \Omega_{\boldsymbol{k},j}}{D}\right)^2\right]^{-1/2}, \end{cases} \tag{C.16}$$

for $j = 1, 2$. The exciton–polariton, i.e., the stationary state of the light–matter interaction, can be described by (C.13) and (C.16).

D Derivation of Equations in Chapter 8

Several key equations in Chap. 8 are derived in this appendix. For the details of basic concepts used in this appendix, the reader is recommended to refer to the related references and papers [D.1–D.14].

D.1 Derivation of (8.1)

Consider a system consisting of charged particles which is localized in a microscopic space. For convenience, it is called a molecule in this section. To avoid confusion, an electric dipole moment is represented by μ, not by $\boldsymbol{p}$. The electric charge, mass, position, velocity, and momentum of the charged particle are represented by e, m, $\boldsymbol{q}$, $\dot{\boldsymbol{q}}$, and $\boldsymbol{p}$, respectively. In order to derive an interaction Hamiltonian of (8.1) for a two-molecule system, the following four approximations are made:

1. In the case when the wavelength of the electromagnetic fields is longer than the size of the molecule, the vector potential $\boldsymbol{A}$ is approximated as a constant irrespective of the position $\boldsymbol{q}$ of the electric charge in the molecule. Under this approximation, the vector potential is represented by the value at the center of gravity $\boldsymbol{R}$ of the molecule, i.e.,

$$\boldsymbol{A}(\boldsymbol{q}) = \boldsymbol{A}(\boldsymbol{R}) \ . \tag{D.1}$$

2. The interaction between the molecule and the magnetic field is neglected because $\boldsymbol{B}$ $(= \nabla \times \boldsymbol{A})$ is zero [see (1)].
3. Only the electric dipole interacts with the electromagnetic fields, i.e., the higher-order multipoles are neglected.
4. The interaction between the molecules due to electron exchange is neglected.

Under these approximations, the Lagrangian L for the two-molecule system is given by [D.1–D.3]

$$L = L_{\mathrm{mol}} + L_{\mathrm{rad}} + L_{\mathrm{int}} \ , \tag{D.2a}$$

where

$$L_{\text{mol}} = \sum_{\zeta} \left[\sum_{\alpha} \frac{m_\alpha \dot{\boldsymbol{q}}_\alpha^2(\zeta)}{2} - V(\zeta) \right] , \tag{D.2b}$$

$$L_{\text{rad}} = \frac{\varepsilon_0}{2} \int \left[\dot{\boldsymbol{A}}^2 - c^2 (\nabla \times \boldsymbol{A})^2 \right] \mathrm{d}^3 r , \tag{D.2c}$$

$$L_{\text{int}} = \sum_{\zeta} \sum_{\alpha} e \dot{\boldsymbol{q}}_\alpha(\zeta) \cdot \boldsymbol{A}(\boldsymbol{R}_\zeta) - V_{\text{inter}} . \tag{D.2d}$$

Greek letters ζ and α are used in (D.2b) and (D.2d) in order to discriminate the two molecules and electric charges, respectively. The Lagrangian L_{mol} of (D.2b) represents the difference between the kinetic energy and potential energy of the Coulomb force. On the other hand, L_{rad} of (D.2c) and L_{int} of (D.2d) represent the electromagnetic field energy in the free space and the interaction energy between the electric charge and the electromagnetic fields, respectively. Using the electric dipole moments $\mu(1)$ and $\mu(2)$ of the two molecules, the intermolecular Coulomb interaction energy V_{inter} in (D.2d) is expressed as [D.3, D.4]

$$V_{\text{inter}} = \frac{1}{4\pi\varepsilon_0 R^3} \left\{ \boldsymbol{\mu}(1) \cdot \boldsymbol{\mu}(2) - 3 \left[\boldsymbol{\mu}(1) \cdot \boldsymbol{e}_R \right] \left[\boldsymbol{\mu}(2) \cdot \boldsymbol{e}_R \right] \right\} , \tag{D.3}$$

where $R\ (= |\boldsymbol{R}| = |\boldsymbol{R}_2 - \boldsymbol{R}_1|)$ is the distance between the two molecules and $\boldsymbol{e}_R$ is a unit vector along $\boldsymbol{R}$.

In order to derive a simple form of the interaction Hamiltonian H_{int}, the original Lagrangian L of (D.2a) is transformed by

$$L_{\text{mult}} = L - \frac{\mathrm{d}}{\mathrm{d}t} \int \boldsymbol{P}^{\perp}(\boldsymbol{r}) \cdot \boldsymbol{A}(\boldsymbol{r}) \mathrm{d}^3 r , \tag{D.4}$$

where $\boldsymbol{P}^{\perp}(\boldsymbol{r})$ is the transverse component of the polarization $\boldsymbol{P}(\boldsymbol{r})$,

$$\begin{aligned} \boldsymbol{P}(\boldsymbol{r}) &= \sum_{\zeta,\alpha} e \left[\boldsymbol{q}_\alpha(\zeta) - \boldsymbol{R}_\zeta \right] \delta(\boldsymbol{r} - \boldsymbol{R}_\zeta) \\ &= \boldsymbol{\mu}(1) \delta(\boldsymbol{r} - \boldsymbol{R}_1) + \boldsymbol{\mu}(2) \delta(\boldsymbol{r} - \boldsymbol{R}_2) . \end{aligned} \tag{D.5}$$

Here, the term ‘transverse’ indicates that the vector under consideration is vertical with respect to the wave vector $\boldsymbol{k}$, i.e., only the transverse photons to the relevant process.

By noting that the transverse component $\boldsymbol{j}^{\perp}(\boldsymbol{r})$ of the current density $\boldsymbol{j}(\boldsymbol{r})$, viz.,

$$\boldsymbol{j}(\boldsymbol{r}) = \sum_{\zeta,\alpha} e \dot{\boldsymbol{q}}_\alpha \delta(\boldsymbol{r} - \boldsymbol{R}_\zeta) , \tag{D.6}$$

satisfies the relation

$$\frac{\mathrm{d}\boldsymbol{P}^{\perp}(\boldsymbol{r})}{\mathrm{d}t} = \boldsymbol{j}^{\perp}(\boldsymbol{r}) , \tag{D.7}$$

equation (D.2d) can be transformed to

$$L_{\text{int}} = \int \boldsymbol{j}^{\perp}(\boldsymbol{r})\cdot\boldsymbol{A}(\boldsymbol{r})\mathrm{d}^3r - V_{\text{inter}} = \int \frac{\mathrm{d}\boldsymbol{P}^{\perp}(\boldsymbol{r})}{\mathrm{d}t}\cdot\boldsymbol{A}(\boldsymbol{r})\mathrm{d}^3r - V_{\text{inter}} , \qquad \text{(D.8)}$$

so that (D.4) reduces to

$$\begin{aligned} L_{\text{mult}} &= L - \int \frac{\mathrm{d}\boldsymbol{P}^{\perp}}{\mathrm{d}t}\cdot\boldsymbol{A}(\boldsymbol{r})\mathrm{d}^3r - \int \boldsymbol{P}^{\perp}(\boldsymbol{r})\cdot\dot{\boldsymbol{A}}(\boldsymbol{r})\mathrm{d}^3r \\ &= L_{\text{mol}} + L_{\text{rad}} - \int \boldsymbol{P}^{\perp}(\boldsymbol{r})\cdot\dot{\boldsymbol{A}}(\boldsymbol{r})\mathrm{d}^3r - V_{\text{inter}} . \end{aligned} \qquad \text{(D.9)}$$

Further, the momenta $\boldsymbol{p}_\alpha$ and $\boldsymbol{\Pi}(\boldsymbol{r})$, which are conjugate to $\boldsymbol{q}_\alpha$ and $\boldsymbol{A}(\boldsymbol{r})$, respectively, are given as

$$\boldsymbol{p}_\alpha = \frac{\partial L_{\text{mult}}}{\partial \dot{\boldsymbol{q}}_\alpha} = \frac{\partial L_{\text{mol}}}{\partial \dot{\boldsymbol{q}}_\alpha} = m_\alpha \dot{\boldsymbol{q}}_\alpha , \qquad \text{(D.10)}$$

and

$$\begin{aligned} \boldsymbol{\Pi}(\boldsymbol{r}) &= \frac{\partial L_{\text{mult}}}{\partial \dot{\boldsymbol{A}}(\boldsymbol{r})} = \frac{\partial L_{\text{rad}}}{\partial \dot{\boldsymbol{A}}(\boldsymbol{r})} - \frac{\partial}{\partial \dot{\boldsymbol{A}}(\boldsymbol{r})} \int \boldsymbol{P}^{\perp}(\boldsymbol{r})\cdot\dot{\boldsymbol{A}}(\boldsymbol{r})\mathrm{d}^3r \\ &= \varepsilon_0 \dot{\boldsymbol{A}}(\boldsymbol{r}) - \boldsymbol{P}^{\perp}(\boldsymbol{r}) = -\varepsilon_0 \boldsymbol{E}^{\perp}(\boldsymbol{r}) - \boldsymbol{P}^{\perp}(\boldsymbol{r}) . \end{aligned} \qquad \text{(D.11)}$$

Using the electric flux density $\boldsymbol{D}(\boldsymbol{r})$ defined by (A.16a) of Appendix A,

$$\boldsymbol{D}(\boldsymbol{r}) = \varepsilon_0 \boldsymbol{E}(\boldsymbol{r}) + \boldsymbol{P}(\boldsymbol{r}) , \qquad \text{(D.12)}$$

the conjugate momentum $\boldsymbol{\Pi}(\boldsymbol{r})$ is written as

$$\boldsymbol{\Pi}(\boldsymbol{r}) = -\boldsymbol{D}^{\perp}(\boldsymbol{r}) . \qquad \text{(D.13)}$$

Thus, by eliminating $\dot{\boldsymbol{q}}_\alpha$ and $\dot{\boldsymbol{A}}(\boldsymbol{r})$ with the help of (D.9)–(D.11), the Hamiltonian H_{mult} is represented by

$$\begin{aligned} H_{\text{mult}} &= \sum_{\zeta,\alpha} \boldsymbol{p}_\alpha(\zeta)\cdot\dot{\boldsymbol{q}}_\alpha(\zeta) + \int \boldsymbol{\Pi}(\boldsymbol{r})\cdot\dot{\boldsymbol{A}}(\boldsymbol{r})\mathrm{d}^3r - L_{\text{mult}} \\ &= \sum_{\zeta} \left[\sum_{\alpha} \frac{\boldsymbol{p}_\alpha^2(\zeta)}{2m_\alpha} + V(\zeta) + \frac{1}{2\varepsilon_0} \int |\boldsymbol{P}_\zeta^{\perp}(\boldsymbol{r})|^2 \mathrm{d}^3r \right] \\ &+ \left\{ \frac{1}{2} \int \left[\frac{\boldsymbol{\Pi}^2(\boldsymbol{r})}{\varepsilon_0} + \varepsilon_0 c^2 \left[\nabla \times \boldsymbol{A}(\boldsymbol{r})\right]^2 \right] \mathrm{d}^3r \right\} \\ &+ \frac{1}{\varepsilon_0} \int \boldsymbol{P}^{\perp}(\boldsymbol{r})\cdot\boldsymbol{\Pi}(\boldsymbol{r})\mathrm{d}^3r , \end{aligned} \qquad \text{(D.14)}$$

where the relation

$$V_{\text{inter}} + \frac{1}{\varepsilon_0}\int \boldsymbol{P}_1^{\perp}(\boldsymbol{r})\cdot\boldsymbol{P}_2^{\perp}(\boldsymbol{r})\mathrm{d}^3r = 0 \tag{D.15}$$

has been used. To see this, note that $\boldsymbol{P}_2(\boldsymbol{r}) = \boldsymbol{P}_2^{\|}(\boldsymbol{r}) + \boldsymbol{P}_2^{\perp}(\boldsymbol{r})$,

$$\int \boldsymbol{P}_1^{\perp}(\boldsymbol{r})\cdot\boldsymbol{P}_2(\boldsymbol{r})\mathrm{d}^3r = \int \boldsymbol{P}_1^{\perp}\cdot\left[\boldsymbol{P}_2^{\|}(\boldsymbol{r}) + \boldsymbol{P}_2^{\perp}(\boldsymbol{r})\right]\mathrm{d}^3r = \int \boldsymbol{P}_1^{\perp}(\boldsymbol{r})\cdot\boldsymbol{P}_2^{\perp}(\boldsymbol{r})\mathrm{d}^3r\,,$$

and use (D.5) that the polarization is expressed as

$$\boldsymbol{P}(\boldsymbol{r}) = \boldsymbol{\mu}(1)\delta(\boldsymbol{r}-\boldsymbol{R}_1) + \boldsymbol{\mu}(2)\delta(\boldsymbol{r}-\boldsymbol{R}_2)\,,$$

to obtain

$$\begin{aligned}
\frac{1}{\varepsilon_0}\int \boldsymbol{P}_1^{\perp}(\boldsymbol{r})\cdot\boldsymbol{P}_2^{\perp}(\boldsymbol{r})\mathrm{d}^3r &= \frac{1}{\varepsilon_0}\int \boldsymbol{P}_1^{\perp}(\boldsymbol{r})\cdot\boldsymbol{P}_2(\boldsymbol{r})\mathrm{d}^3r \\
&= \frac{\mu_i(1)\mu_j(2)}{\varepsilon_0}\int \delta_{ij}^{\perp}(\boldsymbol{r}-\boldsymbol{R}_1)\cdot\delta(\boldsymbol{r}-\boldsymbol{R}_2)\mathrm{d}^3r \\
&= \frac{\mu_i(1)\mu_j(2)}{\varepsilon_0}\delta_{ij}^{\perp}(\boldsymbol{R}_1-\boldsymbol{R}_2) \\
&= \frac{1}{4\pi\varepsilon_0 R^3}\mu_i(1)\mu_j(2)(3\hat{e}_{Ri}\hat{e}_{Rj}-\delta_{ij}) \\
&= \frac{1}{4\pi\varepsilon_0 R^3}\left\{3\left[\boldsymbol{\mu}(1)\cdot\boldsymbol{e}_R\right]\left[\boldsymbol{\mu}(2)\cdot\boldsymbol{e}_R\right] - \boldsymbol{\mu}(1)\cdot\boldsymbol{\mu}(2)\right\}.
\end{aligned}$$

This is equal to (D.3) except that the sign is opposite. Thus, the right-hand side of (D.15) is found to be zero.

Returning to (D.14), the first and second terms on the right-hand side represent the motion of charged particles in each molecule and the electromagnetic fields in the free space, respectively. The third term represents the interaction between the charged particles and the electromagnetic fields. A Hamiltonian of the form given by (D.14) is called a multipolar Hamiltonian because the polarization $\boldsymbol{P}(\boldsymbol{r})$ on the right-hand side of (D.14) can be expressed by a dipole, quadrupole, ..., and the 2^l th pole moment ($l = 1, 2, 3, \ldots$).

The last term of (D.14) can be transformed by using (D.5) and (D.13), and is expressed as

$$\begin{aligned}
\frac{1}{\varepsilon_0}\int \boldsymbol{P}^{\perp}(\boldsymbol{r})\cdot\boldsymbol{\Pi}(\boldsymbol{r})\mathrm{d}^3r &= -\frac{1}{\varepsilon_0}\int \boldsymbol{P}^{\perp}(\boldsymbol{r})\cdot\boldsymbol{D}^{\perp}(\boldsymbol{r})\mathrm{d}^3r \\
&= -\frac{1}{\varepsilon_0}\int \boldsymbol{P}(\boldsymbol{r})\cdot\boldsymbol{D}^{\perp}(\boldsymbol{r})\mathrm{d}^3r \\
&= -\frac{1}{\varepsilon_0}\left[\boldsymbol{\mu}(1)\cdot\boldsymbol{D}^{\perp}(\boldsymbol{R}_1) + \boldsymbol{\mu}(2)\cdot\boldsymbol{D}^{\perp}(\boldsymbol{R}_2)\right].
\end{aligned} \tag{D.16}$$

Since the system under study is quantized, the quantities $\boldsymbol{\mu}(1)$, $\boldsymbol{\mu}(2)$, $\boldsymbol{D}^{\perp}(\boldsymbol{R}_1)$, and $\boldsymbol{D}^{\perp}(\boldsymbol{R}_2)$ of (D.16) must be replaced by the quantum mechanical operators $\hat{\boldsymbol{\mu}}(1)$, $\hat{\boldsymbol{\mu}}(2)$, $\hat{\boldsymbol{D}}^{\perp}(\boldsymbol{R}_1)$, and $\hat{\boldsymbol{D}}^{\perp}(\boldsymbol{R}_2)$. Then the second line of (D.16) can be regarded as an interaction Hamiltonian operator $\hat{V}$, which is expressed as

$$\hat{V} = -\frac{1}{\varepsilon_0}\left[\hat{\boldsymbol{\mu}}(1)\cdot\hat{\boldsymbol{D}}^{\perp}(\boldsymbol{R}_1) + \hat{\boldsymbol{\mu}}(2)\cdot\hat{\boldsymbol{D}}^{\perp}(\boldsymbol{R}_2)\right] . \tag{D.17}$$

From the approximation (1) mentioned at the beginning of this section, $\hat{\boldsymbol{D}}^{\perp}(\boldsymbol{R}_1)$ and $\hat{\boldsymbol{D}}^{\perp}(\boldsymbol{R}_2)$ can be replaced by the electric flux density operators $\hat{\boldsymbol{D}}(\boldsymbol{r}_{\mathrm{S}})$ and $\hat{\boldsymbol{D}}(\boldsymbol{r}_{\mathrm{P}})$ at the positions $\boldsymbol{r}_{\mathrm{S}}$ and $\boldsymbol{r}_{\mathrm{P}}$ of electric charges, respectively. Further, by representing $\hat{\boldsymbol{\mu}}(1)$ and $\hat{\boldsymbol{\mu}}(2)$ as $\boldsymbol{p}_{\mathrm{S}}$ and $\boldsymbol{p}_{\mathrm{P}}$, respectively, (D.17) can be reduced to (8.1).

Equation (8.1), i.e., representation by the multipolar Hamiltonian, has several advantages:

- it can describe the molecular interaction induced via a photon of the transverse wave, because the static Coulomb interaction is not included,
- retardation effect is included,
- the Hamiltonian is expressed in a simple form.

D.2 Derivation of (8.2)

In the quantum theory of the electromagnetic field, the vector potential, its conjugate momentum, the electric field, and the electric flux density are quantum mechanical operators $\hat{\boldsymbol{A}}$, $\hat{\boldsymbol{\Pi}}$, $\hat{\boldsymbol{E}}$, and $\hat{\boldsymbol{D}}$, respectively. Among them, $\hat{\boldsymbol{A}}$ and $\hat{\boldsymbol{\Pi}}$ are represented by analogy with the mode expansion in the classical theory as

$$\hat{\boldsymbol{A}}(\boldsymbol{r}) = \sum_{\boldsymbol{k}}\sum_{\lambda=1}^{2}\left(\frac{\hbar}{2\varepsilon_0 V\omega_{\boldsymbol{k}}}\right)^{1/2}\boldsymbol{e}_{\lambda}(\boldsymbol{k})\left[\hat{a}_{\lambda}(\boldsymbol{k})\mathrm{e}^{\mathrm{i}\boldsymbol{k}\cdot\boldsymbol{r}} + \hat{a}_{\lambda}^{\dagger}(\boldsymbol{k})\mathrm{e}^{-\mathrm{i}\boldsymbol{k}\cdot\boldsymbol{r}}\right] , \tag{D.18}$$

and

$$\begin{aligned}\hat{\boldsymbol{\Pi}}(\boldsymbol{r}) &= \varepsilon_0\frac{\partial\hat{\boldsymbol{A}}}{\partial t} \\ &= -\mathrm{i}\sum_{\boldsymbol{k}}\sum_{\lambda=1}^{2}\left(\frac{\hbar\varepsilon_0\omega_{\boldsymbol{k}}}{2V}\right)^{1/2}\boldsymbol{e}_{\lambda}(\boldsymbol{k})\left[\hat{a}_{\lambda}(\boldsymbol{k})\mathrm{e}^{\mathrm{i}\boldsymbol{k}\cdot\boldsymbol{r}} - \hat{a}_{\lambda}^{\dagger}(\boldsymbol{k})\mathrm{e}^{-\mathrm{i}\boldsymbol{k}\cdot\boldsymbol{r}}\right] ,\end{aligned} \tag{D.19}$$

respectively. Since the relation between the electric field $\hat{\boldsymbol{E}}$ and the electric flux density $\hat{\boldsymbol{D}}$ is expressed by using the polarization $\hat{\boldsymbol{P}}$ as [see (A.16a) of Appendix A]

$$\hat{\boldsymbol{D}}(\boldsymbol{r}) = \varepsilon_0\hat{\boldsymbol{E}}(\boldsymbol{r}) + \hat{\boldsymbol{P}}(\boldsymbol{r}) , \tag{D.20}$$

the relation between the conjugate momentum $\hat{\boldsymbol{\Pi}}$ and the electric flux density $\hat{\boldsymbol{D}}$ for the Hamiltonian of (D.14) is given from (D.11) and (D.13) as

$$\hat{\boldsymbol{\Pi}}(\boldsymbol{r}) = \varepsilon_0\frac{\partial\hat{\boldsymbol{A}}}{\partial t} - \hat{\boldsymbol{P}}(\boldsymbol{r}) = -\varepsilon_0\hat{\boldsymbol{E}}^{\perp}(\boldsymbol{r}) - \hat{\boldsymbol{P}}^{\perp}(\boldsymbol{r}) = -\hat{\boldsymbol{D}}^{\perp}(\boldsymbol{r}) . \tag{D.21}$$

Thus, using (D.19) and (D.21), one obtains

$$\hat{\boldsymbol{D}}(\boldsymbol{r}) = \sum_{\boldsymbol{k}} \sum_{\lambda=1}^{2} \mathrm{i} \left(\frac{\varepsilon_0 \hbar \omega_{\boldsymbol{k}}}{2V} \right)^{1/2} \boldsymbol{e}_\lambda(\boldsymbol{k}) \left[\hat{a}_\lambda(\boldsymbol{k}) \mathrm{e}^{\mathrm{i}\boldsymbol{k}\cdot\boldsymbol{r}} - \hat{a}_\lambda^\dagger(\boldsymbol{k}) \mathrm{e}^{-\mathrm{i}\boldsymbol{k}\cdot\boldsymbol{r}} \right] ,$$

which is identical to(8.2).

D.3 Derivation of (8.3)

Substituting (8.2) and an electric dipole operator $\hat{\boldsymbol{p}}_\alpha = [\hat{B}(\boldsymbol{r}_\alpha) + \hat{B}^\dagger(\boldsymbol{r}_\alpha)]\boldsymbol{p}_\alpha$ into (8.1), one derives

$$\hat{V} = -\frac{\mathrm{i}}{\varepsilon_0} \sum_{\alpha=\mathrm{S}}^{\mathrm{P}} \sum_{\boldsymbol{k}} \sum_{\lambda=1}^{2} \left[\hat{B}(\boldsymbol{r}_\alpha) + \hat{B}^\dagger(\boldsymbol{r}_\alpha) \right] \boldsymbol{p}_\alpha \cdot \boldsymbol{e}_\lambda(\boldsymbol{k}) \left(\frac{\varepsilon_0 \hbar \omega_{\boldsymbol{k}}}{2V} \right)^{1/2} . \quad \text{(D.22)}$$

In order to transform this equation, we use the following relations between annihilation and creation operators for photons $(\hat{a}_\lambda(\boldsymbol{k}), \hat{a}_\lambda^\dagger(\boldsymbol{k}))$ and exciton–polaritons $(\hat{\xi}(\boldsymbol{k}), \hat{\xi}^\dagger(\boldsymbol{k}))$ as

$$\hat{a}_\lambda(\boldsymbol{k}) = w_\lambda(\boldsymbol{k})\hat{\xi}(\boldsymbol{k}) - y_\lambda(\boldsymbol{k})\hat{\xi}^\dagger(-\boldsymbol{k}) , \quad \text{(D.23a)}$$

$$\hat{a}_\lambda^\dagger(\boldsymbol{k}) = w_\lambda(\boldsymbol{k})\hat{\xi}^\dagger(\boldsymbol{k}) - y_\lambda(\boldsymbol{k})\hat{\xi}(-\boldsymbol{k}) , \quad \text{(D.23b)}$$

$$\hat{b}_\lambda(\boldsymbol{k}) = X_\lambda(\boldsymbol{k})\hat{\xi}(\boldsymbol{k}) - Z_\lambda(\boldsymbol{k})\hat{\xi}^\dagger(-\boldsymbol{k}) , \quad \text{(D.23c)}$$

$$\hat{b}_\lambda^\dagger(\boldsymbol{k}) = X_\lambda(\boldsymbol{k})\hat{\xi}^\dagger(\boldsymbol{k}) - Z_\lambda(\boldsymbol{k})\hat{\xi}(-\boldsymbol{k}) . \quad \text{(D.23d)}$$

The relations represented by (D.23a–d) correspond to (C.6) of Appendix C. However, the creation and the annihilation operators of exciton–polaritons are both required here because the rotating wave approximation is not employed. Elements of the matrix U in (C.6) correspond to the coefficients $w_\lambda(\boldsymbol{k})$, $y_\lambda(\boldsymbol{k})$, $X_\lambda(\boldsymbol{k})$, and $Z_\lambda(\boldsymbol{k})$. By deriving the inverse matrix of U, the creation and annihilation operators of the exciton–polariton can be expressed in terms of those of the photon and exciton as

$$\hat{\xi}^\dagger(\boldsymbol{k}) = \sum_\lambda \left[w_\lambda(\boldsymbol{k})\hat{a}^\dagger(\boldsymbol{k}) + y_\lambda(\boldsymbol{k})\hat{a}_\lambda(-\boldsymbol{k}) \right] + \sum_{\lambda'} \left[X_{\lambda'}(\boldsymbol{k})\hat{b}_{\lambda'}^\dagger(\boldsymbol{k}) + Z_{\lambda'}(\boldsymbol{k})\hat{b}_{\lambda'}(-\boldsymbol{k}) \right] , \quad \text{(D.24a)}$$

$$\hat{\xi}(\boldsymbol{k}) = \sum_\lambda \left[w_\lambda(\boldsymbol{k})\hat{a}_\lambda(\boldsymbol{k}) + y_\lambda(\boldsymbol{k})\hat{a}_\lambda^\dagger(-\boldsymbol{k}) \right] + \sum_{\lambda'} \left[X_{\lambda'}(\boldsymbol{k})\hat{b}_{\lambda'}(\boldsymbol{k}) + Z_{\lambda'}(\boldsymbol{k})\hat{b}_{\lambda'}^\dagger(-\boldsymbol{k}) \right] , \quad \text{(D.24b)}$$

respectively. The coefficients $w_\lambda(\boldsymbol{k})$, $y_\lambda(\boldsymbol{k})$, $X_\lambda(\boldsymbol{k})$, and $Z_\lambda(\boldsymbol{k})$ are derived by solving a secular equation based on (D.24a) and (D.24b), as shown in (C.10) of Appendix C. The results are [D.5–D.9]

$$w_\lambda(\boldsymbol{k}) = \frac{\Omega(k)+\omega_{\boldsymbol{k}}}{2\sqrt{\Omega(k)\omega_{\boldsymbol{k}}}} \frac{\Omega^2(k)-\Omega^2}{\sqrt{\left[\Omega^2(k)-\Omega^2\right]^2 + \left[\Omega^2(k)-\omega_{\boldsymbol{k}}^2\right]\left[\Omega^2(k)-\Omega^2\right]}} , \tag{D.25a}$$

$$y_\lambda(\boldsymbol{k}) = -\frac{\Omega(k)-\omega_{\boldsymbol{k}}}{\Omega(k)+\omega_{\boldsymbol{k}}} w_\lambda(\boldsymbol{k}) . \tag{D.25b}$$

Thus it follows

$$w_\lambda(\boldsymbol{k}) + y_\lambda(\boldsymbol{k}) = \sqrt{\frac{\omega_{\boldsymbol{k}}}{\Omega(k)}} \frac{\Omega^2(k)-\Omega^2}{\sqrt{\left[\Omega^2(k)-\Omega^2\right]^2 + \left[\Omega^2(k)-\omega_{\boldsymbol{k}}^2\right]\left[\Omega^2(k)-\Omega^2\right]}} . \tag{D.26}$$

On the other hand, (D.22) is transformed to

$$\begin{aligned} \hat{V} = -\mathrm{i} \sum_{\alpha=\mathrm{S}}^{\mathrm{P}} \sum_{\boldsymbol{k}} \sum_{\lambda=1}^{2} & \left[\hat{B}(\boldsymbol{r}_\alpha) + \hat{B}^\dagger(\boldsymbol{r}_\alpha)\right] \left[\boldsymbol{p}_\alpha \cdot \boldsymbol{e}_\lambda(\boldsymbol{k})\right] \sqrt{\frac{\hbar\omega_{\boldsymbol{k}}}{2\varepsilon_0 V}} \\ & \times \Big\{ [w_\lambda(\boldsymbol{k})\hat{\xi}(\boldsymbol{k}) - y_\lambda(\boldsymbol{k})\hat{\xi}^\dagger(-\boldsymbol{k})] \mathrm{e}^{\mathrm{i}\boldsymbol{k}\cdot\boldsymbol{r}_\alpha} \\ & \quad - [w_\lambda(\boldsymbol{k})\hat{\xi}^\dagger(\boldsymbol{k}) - y_\lambda(\boldsymbol{k})\hat{\xi}(-\boldsymbol{k})] \mathrm{e}^{-\mathrm{i}\boldsymbol{k}\cdot\boldsymbol{r}_\alpha} \Big\} , \end{aligned} \tag{D.27}$$

by using (D.23a) and (D.23b). Swapping $-\boldsymbol{k}$ and $\boldsymbol{k}$ in { } in (D.27) and using the relations $w_\lambda(\boldsymbol{k}) = w_\lambda(-\boldsymbol{k})$, $y_\lambda(\boldsymbol{k}) = y_\lambda(-\boldsymbol{k})$, $X_\lambda(\boldsymbol{k}) = X_\lambda(-\boldsymbol{k})$, and $Z_\lambda(\boldsymbol{k}) = Z_\lambda(-\boldsymbol{k})$, (D.27) can be rewritten as

$$\begin{aligned} \hat{V} = -\mathrm{i} \sum_{\alpha=\mathrm{S}}^{\mathrm{P}} \sum_{\boldsymbol{k}} \sum_{\lambda=1}^{2} & \left(\frac{\hbar\omega_{\boldsymbol{k}}}{2\varepsilon_0 V}\right)^{1/2} \left[\hat{B}(\boldsymbol{r}_\alpha) + \hat{B}^\dagger(\boldsymbol{r}_\alpha)\right] \left[\boldsymbol{p}_\alpha \cdot \boldsymbol{e}_\lambda(\boldsymbol{k})\right] \\ & \times \Big\{ [w_\lambda(\boldsymbol{k}) + y_\lambda(\boldsymbol{k})][\hat{\xi}(\boldsymbol{k})\mathrm{e}^{\mathrm{i}\boldsymbol{k}\cdot\boldsymbol{r}_\alpha} - \hat{\xi}^\dagger(\boldsymbol{k})\mathrm{e}^{-\mathrm{i}\boldsymbol{k}\cdot\boldsymbol{r}_\alpha}] \Big\} , \end{aligned} \tag{D.28}$$

with $\omega_{\boldsymbol{k}} = ck$. Substituting (D.26) into this equation, one has

$$\begin{aligned} \hat{V} = -\mathrm{i} \sum_{\alpha=s}^{p} \sum_{\boldsymbol{k}} \sum_{\lambda=1}^{2} & \sqrt{\frac{\hbar}{2\varepsilon_0 V}} \left[\hat{B}(\boldsymbol{r}_\alpha) + \hat{B}^\dagger(\boldsymbol{r}_\alpha)\right] \left[\boldsymbol{p}_\alpha \cdot \boldsymbol{e}_\lambda(\boldsymbol{k})\right] \\ & \times \frac{ck}{\sqrt{\Omega(k)}} \frac{\Omega^2(k)-\Omega^2}{\sqrt{\left[\Omega^2(k)-\Omega^2\right]^2 + \left[\Omega^2(k)-(ck)^2\right]\left[\Omega^2(k)-\Omega^2\right]}} . \end{aligned} \tag{D.29}$$

Further, substituting the following relations into (D.29),

$$\begin{aligned} K_\alpha(\boldsymbol{k}) &= \sum_{\lambda=1}^{2} \left[\boldsymbol{p}_\alpha \cdot \boldsymbol{e}_\lambda(\boldsymbol{k})\right] f(k) \mathrm{e}^{\mathrm{i}\boldsymbol{k}\cdot\boldsymbol{r}_\alpha} , \\ K_\alpha^*(\boldsymbol{k}) &= \sum_{\lambda=1}^{2} \left[\boldsymbol{p}_\alpha \cdot \boldsymbol{e}_\lambda(\boldsymbol{k})\right] f(k) \mathrm{e}^{-\mathrm{i}\boldsymbol{k}\cdot\boldsymbol{r}_\alpha} , \end{aligned} \tag{D.30}$$

and

$$f(k) = \frac{ck}{\sqrt{\Omega(k)}} \frac{\Omega^2(k) - \Omega^2}{\sqrt{\left[\Omega^2(k) - \Omega^2\right]^2 + \left[\Omega^2(k) - \Omega^2\right]\left[\Omega^2(k) - (ck)^2\right]}}$$
$$= \frac{ck}{\sqrt{\Omega(k)}} \sqrt{\frac{\Omega^2(k) - \Omega^2}{2\Omega^2(k) - (ck)^2 - \Omega^2}} , \tag{D.31}$$

which correspond to (8.4a) and (8.4b), one finally obtains

$$\hat{V} = -\mathrm{i}\sqrt{\frac{\hbar}{2\varepsilon_0 V}} \sum_{\alpha=\mathrm{S}}^{\mathrm{P}} \left[\hat{B}(\boldsymbol{r}_\alpha) + \hat{B}^\dagger(\boldsymbol{r}_\alpha)\right] \sum_{\boldsymbol{k}} \left[K_\alpha(\boldsymbol{k})\hat{\xi}(\boldsymbol{k}) - K_\alpha^*(\boldsymbol{k})\hat{\xi}^\dagger(\boldsymbol{k})\right] , \tag{D.32}$$

which is none other than (8.3). For more details about the derivation of (8.3), the reader is referred to [D.5–D.9].

D.4 Projection Operator Method and Derivation of (8.5)

D.4.1 Definition of a Projection Operator

When a material system interacts with electromagnetic fields, the total Hamiltonian $\hat{H}$ can be expressed as the sum of the Hamiltonian for the isolated system $\hat{H}_0$ and the interaction Hamiltonian $\hat{V}$, viz.,

$$\hat{H} = \hat{H}_0 + \hat{V} . \tag{D.33}$$

Eigenstates $|\psi_j\rangle$ of $\hat{H}$ satisfy the following eigenequation with eigenvalues E_j as

$$\hat{H}|\Psi_j\rangle = E_j|\Psi_j\rangle . \tag{D.34}$$

The eigenstate of $\hat{H}_0$ is represented by $|\phi_j\rangle$, which corresponds to $|\phi_1\rangle$ and $|\phi_2\rangle$ of Sect. 8.1. Using these eigenstates, the operator P is defined as

$$P = \sum_{j=1}^{N} |\phi_j\rangle\langle\phi_j| , \tag{D.35}$$

which is called a projection operator. In Sect. 8.2, the projection operator was defined by fixing $N = 2$, i.e., by using two states $|\phi_1\rangle$ and $|\phi_2\rangle$. Applying this operator to an arbitrary state $|\Psi\rangle$, one derives

$$P|\Psi\rangle = \sum_{j=1}^{N} |\phi_j\rangle\langle\phi_j|\Psi\rangle . \tag{D.36}$$

Since the inner product $\langle\phi_j|\Psi\rangle$ is a constant denoted by c_j, the right-hand side of this equation is expressed as $\sum_{j=1}^{N} c_j|\phi_j\rangle$. This means that the operator P projects the arbitrary state $|\Psi\rangle$ onto the state vector space (P-space) composed of the states $|\phi_j\rangle$ ($j = 1, \ldots, N$). The reason why the P-space is composed of only two states $|\phi_1\rangle$ and $|\phi_2\rangle$ is to make the discussion as simple as possible by limiting the number of energy eigenstates of the sample and probe in the subsystem (N). The states of the subsystem (M) are taken into account implicitly by the form of the effective operator.

Using the operator P, one can derive an operator $\hat{O}_{\text{eff}}$ in the P-space which is used to calculate the expectation value $\langle\Psi|\hat{O}|\Psi\rangle$ of the arbitrary operator $\hat{O}$ and state $|\Psi\rangle$, i.e., to represent it in the form $\langle\phi_i|\hat{O}_{\text{eff}}|\phi_j\rangle$. This operator $\hat{O}_{\text{eff}}$ is called an effective operator because it is equivalent to the original operator $\hat{O}$ as long as it is used in the P-space [D.9–D.14]. Several characteristics of the operator P are listed in the following as a preparation for deriving the effective operator $\hat{O}_{\text{eff}}$ in Sect. D.4.2.

Since $|\phi_j\rangle$ is orthonormal, the relations

$$P = P^{\dagger}\,, \quad P^2 = P \tag{D.37}$$

hold, where $P^{\dagger}$ is the Hermitian conjugate operator of P. Since $P = P^{\dagger}$ [see (D.37)], P is a Hermitian operator.

The projection operator onto the complementary space of P (Q-space) is given by

$$Q = 1 - P\,, \tag{D.38}$$

and the relations

$$Q = Q^{\dagger}\,, \quad Q^2 = Q \tag{D.39}$$

then hold. Further, since any states in the P-space are mutually orthogonal to any states in the Q-space, the relation

$$PQ = QP = 0 \tag{D.40}$$

must hold. Since $|\phi_j\rangle$ is an eigenstate of $\hat{H}_0$, the commutator of the projection operator and $\hat{H}_0$ vanishes:

$$[P, \hat{H}_0] = P\hat{H}_0 - \hat{H}_0 P\,, \quad [Q, \hat{H}_0] = Q\hat{H}_0 - \hat{H}_0 Q = 0 \tag{D.41}$$

D.4.2 Derivation of an Effective Operator

This section is concerned with the effective operator $\hat{O}_{\text{eff}}$ derived by using the operators P and Q. Since an arbitrary state $|\Psi\rangle$ is expressed as a linear superposition of the eigenstates $|\Psi_j\rangle$ of $\hat{H}$, the following discussions are made not for $|\Psi\rangle$ but for $|\Psi_j\rangle$.

Using the eigenstate $|\Psi_j\rangle$, the states $|\Psi_j^{(1)}\rangle$ and $|\Psi_j^{(2)}\rangle$ in the P- and Q-spaces, respectively, are defined by

$$|\Psi_j^{(1)}\rangle = P|\Psi_j\rangle\,, \quad |\Psi_j^{(2)}\rangle = Q|\Psi_j\rangle\,. \tag{D.42}$$

Noting that $P + Q = 1$, one derives

$$|\Psi_j\rangle = (P+Q)|\Psi_j\rangle = P|\Psi_j\rangle + Q|\Psi_j\rangle = |\Psi_j^{(1)}\rangle + |\Psi_j^{(2)}\rangle\,. \tag{D.43}$$

On the other hand, using the relations $P^2 = P$ and $Q^2 = Q$ [see (D.37) and (D.39)], one finds

$$P|\Psi_j^{(1)}\rangle = PP|\Psi_j\rangle = P|\Psi_j\rangle = |\Psi_j^{(1)}\rangle\,, \tag{D.44a}$$

$$Q|\psi_j^{(2)}\rangle = QQ|\psi_j\rangle = Q|\psi_j\rangle = |\psi_j^{(2)}\rangle\,. \tag{D.44b}$$

Substituting (D.44a) and (D.44b) into (D.43), one has

$$|\Psi_j\rangle = P|\Psi_j^{(1)}\rangle + Q|\Psi_j^{(2)}\rangle\,. \tag{D.45}$$

Since the relation $(E_j - \hat{H}_0)|\Psi_j\rangle = \hat{V}|\Psi_j\rangle$ holds due to (D.33) and (D.34), substitution of (D.45) into this relation leads to

$$(E_j - \hat{H}_0)P|\Psi_j^{(1)}\rangle + (E_j - \hat{H}_0)Q|\Psi_j^{(2)}\rangle = \hat{V}P|\Psi_j^{(1)}\rangle + \hat{V}Q|\Psi_j^{(2)}\rangle\,. \tag{D.46}$$

Applying the operator P from the left on (D.46) and using (D.40) and the relation $P^2 = P$, we obtain

$$(E_j - \hat{H}_0)P|\Psi_j^{(1)}\rangle = P\hat{V}P|\Psi_j^{(1)}\rangle + P\hat{V}Q|\Psi_j^{(2)}\rangle\,. \tag{D.47}$$

Similarly, applying the operator Q from the left on (D.46) and using (D.40) and the relation $Q^2 = Q$, one derives

$$(E_j - \hat{H}_0)Q|\Psi_j^{(2)}\rangle = Q\hat{V}P|\Psi_j^{(1)}\rangle + Q\hat{V}Q|\Psi_j^{(2)}\rangle\,. \tag{D.48}$$

Moving the second term on the right-hand side to the left-hand side, (D.48) can be transformed to

$$\begin{aligned} Q|\Psi_j^{(2)}\rangle &= (E_j - \hat{H}_0 - Q\hat{V})^{-1}Q\hat{V}P|\Psi_j^{(1)}\rangle \\ &= \left\{(E_j - \hat{H}_0)[1 - (E_j - \hat{H}_0)^{-1}Q\hat{V}]\right\}^{-1} Q\hat{V}P|\Psi_j^{(1)}\rangle \\ &= \hat{J}(E_j - \hat{H}_0)^{-1}Q\hat{V}P|\Psi_j^{(1)}\rangle\,, \end{aligned} \tag{D.49}$$

where

$$\hat{J} = \left[1 - (E_j - \hat{H}_0)^{-1}Q\hat{V}\right]^{-1}\,. \tag{D.50}$$

Equation (D.49) shows that $Q|\Psi_j^{(2)}\rangle$ is formally expressed by $|\Psi_j^{(1)}\rangle$.

Substitution of (D.49) into (D.47) gives

$$\begin{aligned} (E_j - \hat{H}_0)P|\Psi_j^{(1)}\rangle &= P\hat{V}P|\Psi_j^{(1)}\rangle + P\hat{V}\hat{J}(E_j - \hat{H}_0)^{-1}Q\hat{V}P|\Psi_j^{(1)}\rangle \\ &= P\hat{V}\hat{J}\left[\hat{J}^{-1} + (E_j - \hat{H}_0)^{-1}Q\hat{V}\right]P|\Psi_j^{(1)}\rangle\,. \end{aligned} \tag{D.51}$$

Since (D.50) leads to

$$\hat{J}^{-1} = 1 - (E_j - \hat{H}_0)^{-1} Q\hat{V} , \tag{D.52}$$

substituting it into the square brackets on the right-hand side of (D.51) we have

$$(E_j - \hat{H}_0)P|\Psi_j^{(1)}\rangle = P\hat{V}\hat{J}P|\Psi_j^{(1)}\rangle . \tag{D.53}$$

This is the equation for $|\Psi_j^{(1)}\rangle$. On the other hand, by substituting (D.49) into the second term on the right-hand side of (D.45), one finds

$$\begin{aligned}|\Psi_j\rangle &= P|\Psi_j^{(1)}\rangle + \hat{J}(E_j - \hat{H}_0)^{-1}Q\hat{V}P|\Psi_j^{(1)}\rangle \\ &= \hat{J}\left[\hat{J}^{-1} + (E_j - \hat{H}_0)^{-1}Q\hat{V}\right]P|\Psi_j^{(1)}\rangle \\ &= \hat{J}P|\Psi_j^{(1)}\rangle ,\end{aligned} \tag{D.54}$$

in which (D.52) was used to derive the last line.

Substituting (D.54) into the normalization condition $\langle\Psi_j|\Psi_j\rangle = 1$, one derives

$$\langle\Psi_j^{(1)}|P\hat{J}^\dagger\hat{J}P|\Psi_j^{(1)}\rangle = 1 . \tag{D.55}$$

This reads

$$\langle\Psi_j^{(1)}|(P\hat{J}^\dagger\hat{J}P)^{1/2}(P\hat{J}^\dagger\hat{J}P)^{1/2}|\Psi_j^{(1)}\rangle = 1 , \tag{D.56}$$

which shows that $|\Psi_j^{(1)}\rangle$ can also be normalized if $(P\hat{J}^\dagger\hat{J}P)^{-1/2}|\Psi_j^{(1)}\rangle$ is defined as $|\Psi_j^{(1)}\rangle$. Therefore, substituting $(P\hat{J}^\dagger\hat{J}P)^{-1/2}|\Psi_j^{(1)}\rangle$ into $|\Psi_j^{(1)}\rangle$ of (D.54), this equation becomes

$$|\Psi_j\rangle = \hat{J}P(P\hat{J}^\dagger\hat{J}P)^{-1/2}|\Psi_j^{(1)}\rangle , \tag{D.57}$$

in which all the states are normalized.

Equation (D.57) expresses $|\Psi_j\rangle$ in terms of $|\Psi_j^{(1)}\rangle$. The effective operator $\hat{O}_{\text{eff}}$ can be derived by using this equation and equating the expectation values of the operators $\hat{O}$ and $\hat{O}_{\text{eff}}$. Hence, the operator $\hat{O}_{\text{eff}}$ satisfying

$$\langle\Psi_i|\hat{O}|\Psi_j\rangle = \langle\Psi_i^{(1)}|\hat{O}_{\text{eff}}|\Psi_j^{(1)}\rangle \tag{D.58}$$

is found by comparing both sides of (D.58) after substituting (D.57) into the left-hand side of (D.58). As a final result, we have

$$\hat{O}_{\text{eff}} = (P\hat{J}^\dagger\hat{J}P)^{-1/2}(P\hat{J}^\dagger\hat{O}\hat{J}P)(P\hat{J}^\dagger\hat{J}P)^{-1/2} . \tag{D.59}$$

Having obtained the effective operator $\hat{O}_{\text{eff}}$ of (D.59), an effective interaction operator $\hat{V}_{\text{eff}}$ in the P-space can be derived if $\hat{O}$ in (D.59) is replaced by $\hat{V}$. It can be written as

$$\hat{V}_{\text{eff}} = (P\hat{J}^\dagger\hat{J}P)^{-1/2}(P\hat{J}^\dagger\hat{V}\hat{J}P)(P\hat{J}^\dagger\hat{J}P)^{-1/2} ,$$

leading to (8.5). As expected, $\hat{V}_{\text{eff}}$ is defined so as to operate only on the states in the P-space.

D.5 Approximation of $\hat{J}$ in (8.5) by $\hat{J}^{(1)}$

In order to obtain an explicit expression for $\hat{J}$, let us consider an operator $[\hat{J}, \hat{H}_0]P$ and apply it to $|\phi_j\rangle$. This yields

$$\begin{aligned}[\hat{J}, \hat{H}_0]P|\Psi_j\rangle &= (\hat{J}\hat{H}_0 - \hat{H}_0\hat{J})P|\Psi_j\rangle \\ &= \big[(E_j - \hat{H}_0)\hat{J} - \hat{J}(E_j - \hat{H}_0)\big]P|\Psi_j\rangle \, . \end{aligned} \tag{D.60}$$

Replacing $(E_j - \hat{H}_0)$ in this equation by $\hat{V}$ with the help of (D.33), (D.34), and (D.54), one derives

$$[\hat{J}, \hat{H}_0]P|\Psi_j\rangle = \hat{V}\hat{J}P|\Psi_j\rangle - \hat{J}(E_j - \hat{H}_0)P|\Psi_j\rangle \, . \tag{D.61}$$

The second term of the right-hand side is transformed to

$$\begin{aligned}(E_j - \hat{H}_0)P|\Psi_j\rangle &= (E_j - \hat{H}_0)P|\Psi_j^{(1)}\rangle \\ &= P\hat{V}P|\Psi_j^{(1)}\rangle + P\hat{V}Q|\Psi_j^{(2)}\rangle \, , \end{aligned} \tag{D.62}$$

by using (D.44a) and (D.47). Further, by substituting (D.49) into $Q|\Psi_j^{(2)}\rangle$ in the second term on the right-hand side, this equation can be rewritten as

$$\begin{aligned}(E_j - \hat{H}_0)P|\Psi_j\rangle &= P\hat{V}P|\Psi_j^{(1)}\rangle + P\hat{V}\hat{J}(E_j - \hat{H}_0)^{-1}Q\hat{V}P|\Psi_j^{(1)}\rangle \\ &= P\hat{V}\hat{J}\big[\hat{J}^{-1} + (E_j - \hat{H}_0)^{-1}Q\hat{V}\big]P|\Psi_j^{(1)}\rangle \, , \end{aligned} \tag{D.63}$$

or, by substituting (D.52) into the right-hand side, this equation reduces to

$$(E_j - \hat{H}_0)P|\Psi_j\rangle = P\hat{V}\hat{J}P|\Psi_j^{(1)}\rangle \, . \tag{D.64}$$

Making use of (D.44a) to rewrite the right-hand side, one finds the relation

$$(E_j - \hat{H}_0)P|\Psi_j\rangle = P\hat{V}\hat{J}P|\Psi_j\rangle \, . \tag{D.65}$$

Substituting this equation into the second term on the right-hand side of (D.61), we finally obtain

$$[\hat{J}, \hat{H}_0]P|\Psi_j\rangle = \hat{V}\hat{J}P|\Psi_j\rangle - \hat{J}P\hat{V}\hat{J}P|\Psi_j\rangle \, . \tag{D.66}$$

From this equation, it follows that

$$[\hat{J}, \hat{H}_0]P = \hat{V}\hat{J}P - \hat{J}P\hat{V}\hat{J}P \, . \tag{D.67}$$

Since this equation is expressed by known operators $\hat{H}_0$, P, and $\hat{V}$, we can determine the operator $\hat{J}$, solving the equation. Here we follow a perturbative method with respect to the magnitude of the interaction. For this derivation, let us assume

$$\hat{J} = \sum_{n=0}^{\infty} g^{(n)} \hat{J}^{(n)} , \tag{D.68}$$

where the nth term $\hat{J}^{(n)}$ contains n $\hat{V}$s and

$$\hat{J}^{(0)} = P . \tag{D.69}$$

Equation (D.69) can be derived by noting that the first term of (D.50) is 1 and $1 = P + Q$ due to (D.38).

We successively obtain $\hat{J}^{(1)}$, $\hat{J}^{(2)}$, ..., $\hat{J}^{(n)}$, substituting (D.68) and (D.69) into (D.67) and equating terms of order $g^{(n)}$ on both sides of (D.67). For $\hat{J}^{(1)}$, as an example, Q is applied from the left to both sides of (D.67) to have

$$Q[\hat{J}^{(1)}, \hat{H}_0]P = Q\hat{V}\hat{J}^{(0)}P - Q\hat{J}^{(0)}P\hat{V}\hat{J}^{(0)}P . \tag{D.70}$$

By substituting (D.69) into this equation, it reads

$$Q[\hat{J}^{(1)}, \hat{H}_0]P = Q\hat{V}P^2 - QP^2\hat{V}P^2 = Q\hat{V}P , \tag{D.71}$$

where the last term was obtained by using (D.37) and (D.40). Taking the matrix element of (D.71) with $\langle \Psi_i|$ and $|\Psi_j\rangle$, we obtain

$$\langle \Psi_i|Q[\hat{J}^{(1)}, \hat{H}_0]P|\Psi_j\rangle = \langle \Psi_i|Q\hat{V}P|\Psi_j\rangle . \tag{D.72}$$

Further, noting that

$$\hat{H}_0 P|\Psi_j\rangle = \hat{H}_0 P|\Psi_j^{(1)}\rangle = P\hat{H}_0|\Psi_j^{(1)}\rangle = PE_P^0|\Psi_j^{(1)}\rangle = E_P^0 P|\Psi_j\rangle \tag{D.73}$$

and

$$\hat{H}_0 Q|\Psi_j\rangle = \hat{H}_0 Q|\Psi_j^{(2)}\rangle = Q\hat{H}_0|\Psi_j^{(2)}\rangle = QE_Q^0|\Psi_j^{(2)}\rangle = E_Q^0 Q|\Psi_j\rangle , \tag{D.74}$$

we can transform the left-hand side of (D.72) to

$$\begin{aligned}\langle \Psi_i|Q(\hat{J}^{(1)}\hat{H}_0 - \hat{H}_0\hat{J}^{(1)})P|\Psi_j\rangle &= \langle \Psi_i|(Q\hat{J}^{(1)}E_P^0 P - QE_Q^0\hat{J}^{(1)}P)|\Psi_j\rangle \\ &= \langle \Psi_i|(Q\hat{J}^{(1)}(E_P^0 - E_Q^0)P)|\Psi_j\rangle . \end{aligned} \tag{D.75}$$

On the other hand, the right-hand side of (D.72) is rewritten as

$$\langle \Psi_i|Q\hat{V}P|\Psi_j\rangle = \langle \Psi_i|Q^2\hat{V}P^2|\Psi_j\rangle , \tag{D.76}$$

by using (D.37) and (D.39). Substituting (D.75) and (D.76) into (D.72) and comparing both sides, one finds

$$Q\hat{J}^{(1)}(E_P^0 - E_Q^0)P = Q^2\hat{V}P^2 . \tag{D.77}$$

Thus, one derives

$$\hat{J}^{(1)} = (E_P^0 - E_Q^0)^{-1}Q\hat{V}P , \tag{D.78}$$

which is proportional to $\hat{V}$. The operator $\hat{J}^{(2)}$ can be obtained similarly. It is proportional to $\hat{V}^2$. The nth term $\hat{J}^{(n)}$ can be derived by repeating this procedure.

D.6 Derivation of (8.9)

We begin with the two states in the P-space

$$\begin{cases} |\phi_1\rangle = |s_{\mathrm{e}}\rangle|p_{\mathrm{g}}\rangle|0_{(\mathrm{M})};\boldsymbol{k},\Omega(k)\rangle\,, \\ |\phi_2\rangle = |s_{\mathrm{g}}\rangle|p_{\mathrm{e}}\rangle|0_{(\mathrm{M})};\boldsymbol{k},\Omega(k)\rangle\,, \end{cases} \tag{D.79}$$

and the effective interaction

$$\hat{V}_{\mathrm{eff}} = 2P\hat{V}Q(E_P^0 - E_Q^0)^{-1}\hat{V}P\,. \tag{D.80}$$

With the help of (D.79) and (D.80), (8.8) is transformed to

$$\begin{aligned} V_{\mathrm{eff}}(ps) = 2\sum_m &\langle 0_{(\mathrm{M})};\boldsymbol{k},\Omega(k)|\langle p_{\mathrm{e}}|\langle s_{\mathrm{g}}|P\hat{V}Q|m\rangle \\ &\times\langle m|Q(E_P^0 - E_Q^0)^{-1}\hat{V}P|s_{\mathrm{e}}\rangle|p_{\mathrm{g}}\rangle|0_{(\mathrm{M})};\boldsymbol{k},\Omega(k)\rangle\,. \end{aligned} \tag{D.81}$$

Note that, as has been shown in (8.3), $\hat{V}$ contains two terms:

- the term composed of $\hat{B}(\boldsymbol{r}_\alpha)$ and $\hat{B}^\dagger(\boldsymbol{r}_\alpha)$ to be applied to $|s\rangle|p\rangle$ of the subsystem (N),
- the term composed of $\hat{\xi}(\boldsymbol{k})$ and $\hat{\xi}^\dagger(\boldsymbol{k})$ to be applied to $|0_{(\mathrm{M})};\boldsymbol{k},\Omega(k)\rangle$ of the subsystem (M).

Furthermore, due to the orthogonality of the state, $V_{\mathrm{eff}}(ps)$ corresponds to nonzero only when the subsystem (M) in the intermediate state $|m\rangle$ is the state $|1_{(\mathrm{M})};\boldsymbol{k},\Omega(k)\rangle$, in which one exciton–polariton exists. Then, since only the state $|s_{\mathrm{g}}\rangle|p_{\mathrm{g}}\rangle|1_{(\mathrm{M})};\boldsymbol{k},\Omega(k)\rangle$ or $|s_{\mathrm{e}}\rangle|p_{\mathrm{e}}\rangle|1_{(\mathrm{M})};\boldsymbol{k},\Omega(k)\rangle$ in the Q-space can contribute to the interaction with a state of the P-space, (D.81) can be rewritten as

$$\begin{aligned} V_{\mathrm{eff}}(ps) = 2\sum_{\boldsymbol{k}} &NK_{\mathrm{P}}(\boldsymbol{k})\langle p_{\mathrm{e}}|\langle s_{\mathrm{g}}|\hat{B}^\dagger(\boldsymbol{r}_{\mathrm{P}})|s_{\mathrm{g}}\rangle|p_{\mathrm{g}}\rangle \\ &\times\langle 0_{(\mathrm{M})};\boldsymbol{k},\Omega(k)|\hat{\xi}(\boldsymbol{k})|1_{(\mathrm{M})};\boldsymbol{k},\Omega(k)\rangle(E_P^0 - E_Q^0)^{-1} \\ &\times(-K_{\mathrm{S}}^*(\boldsymbol{k})N)\langle p_{\mathrm{g}}|\langle s_{\mathrm{g}}|\hat{B}(\boldsymbol{r}_{\mathrm{S}})|s_{\mathrm{e}}\rangle|p_{\mathrm{g}}\rangle \\ &\times\langle 1_{(\mathrm{M})};\boldsymbol{k},\Omega(k)|\hat{\xi}^\dagger(\boldsymbol{k})|0_{(\mathrm{M})};\boldsymbol{k},\Omega(k)\rangle \\ +2\sum_{\boldsymbol{k}} &NK_{\mathrm{S}}(\boldsymbol{k})\langle p_{\mathrm{e}}|\langle s_{\mathrm{g}}|\hat{B}(\boldsymbol{r}_{\mathrm{S}})|s_{\mathrm{e}}\rangle|p_{\mathrm{e}}\rangle \\ &\times\langle 0_{(\mathrm{M})};\boldsymbol{k},\Omega(k)|\hat{\xi}(\boldsymbol{k})|1_{(\mathrm{M})};\boldsymbol{k},\Omega(k)\rangle(E_P^0 - E_Q^0)^{-1} \\ &\times(-K_{\mathrm{P}}^*(\boldsymbol{k})N)\langle p_{\mathrm{e}}|\langle s_{\mathrm{e}}|\hat{B}^\dagger(\boldsymbol{r}_{\mathrm{P}})|s_{\mathrm{e}}\rangle|p_{\mathrm{g}}\rangle \\ &\times\langle 1_{(\mathrm{M})};\boldsymbol{k},\Omega(k)|\hat{\xi}^\dagger(\boldsymbol{k})|0_{(\mathrm{M})};\boldsymbol{k},\Omega(k)\rangle\,, \end{aligned} \tag{D.82}$$

where N is derived from (8.3) and given by

$$N = -\mathrm{i}\left(\frac{\hbar}{2\varepsilon_0 V}\right)^{1/2}\,. \tag{D.83}$$

Noting that the state vectors are normalized, i.e.,

$$\langle s_g|s_g\rangle = \langle p_g|p_g\rangle = \langle p_e|p_e\rangle = \langle s_e|s_e\rangle = 1\,,$$

$$\langle p_e|\hat{B}^{\dagger}(\boldsymbol{r}_P)|p_g\rangle = \langle s_g|\hat{B}(\boldsymbol{r}_S)|s_e\rangle = 1\,,$$

$$\langle 0_{(M)};\boldsymbol{k},\Omega(k)|\hat{\xi}(\boldsymbol{k})|1_{(M)};\boldsymbol{k},\Omega(k)\rangle = \langle 1_{(M)};\boldsymbol{k},\Omega(k)|\hat{\xi}(\boldsymbol{k})^{\dagger}|1_{(M)};\boldsymbol{k},\Omega(k)\rangle = 1\,, \tag{D.84}$$

we can express $V_{\text{eff}}(ps)$ in the following form:

$$\begin{aligned} V_{\text{eff}}(ps) &= 2\sum_{\boldsymbol{k}} \frac{[NK_P(\boldsymbol{k})][-K_S^*(\boldsymbol{k})N]}{[E(s^*)+E(p)]-[E(s)+E(p)+\hbar\Omega(k)]} \\ &\quad +2\sum_{\boldsymbol{k}} \frac{[NK_S(\boldsymbol{k})][-K_P^*(\boldsymbol{k})N]}{[E(s^*)+E(p)]-[E(s^*)+E(p^*)+\hbar\Omega(k)]} \\ &= \sum_{\boldsymbol{k}} \frac{2N^2}{\hbar}\left[\frac{K_P(\boldsymbol{k})K_S^*(\boldsymbol{k})}{\Omega(k)-\Omega_0(s)} + \frac{K_S(\boldsymbol{k})K_P^*(\boldsymbol{k})}{\Omega(k)+\Omega_0(p)}\right]. \end{aligned} \tag{D.85}$$

The ground and excited eigenenergies of the state $|s\rangle$ are expressed as $E(s)$ and $E(s^*)$, respectively, in this equation in order to calculate the values of E_P^0 or E_Q^0. On the other hand, those for the $|p\rangle$ state are expressed as $E(p)$ and $E(p^*)$, respectively. Differences between the excited and ground state energies are expressed as $\hbar\Omega_0(s) = E(s^*) - E(s)$ and $\hbar\Omega_0(p) = E(p^*) - E(p)$. Substitution of (D.83) and usual transformation from $\sum_{\boldsymbol{k}}$ to $\int \mathrm{d}^3k$ gives (D.85)

$$V_{\text{eff}}(ps) = -\frac{1}{(2\pi)^3\varepsilon_0}\int \mathrm{d}^3k \left[\frac{K_P(\boldsymbol{k})K_S^*(\boldsymbol{k})}{\Omega(k)-\Omega_0(s)} + \frac{K_S(\boldsymbol{k})K_P^*(\boldsymbol{k})}{\Omega(k)+\Omega_0(p)}\right], \tag{D.86}$$

which is none other than (8.9).

D.7 Derivation of (8.12)

One should note the three points (a)–(c) given in Sect. 8.2 in order to integrate (8.9). That is, (8.10a), (8.10b), and (8.11) are used. Further, with respect to $K_P(\boldsymbol{k})$ and $K_S(\boldsymbol{k})$, all the terms in (8.4) except for the term $\mathrm{e}^{\mathrm{i}\boldsymbol{k}\cdot\boldsymbol{r}_\alpha}$ are approximated as constants because $f(k)$ of (8.11) is nearly constant. (It has been confirmed by more detailed analyses that the results given in this section can be derived without using this approximation [D.9].) Then (8.9) becomes

$$
\begin{aligned}
V_{\mathrm{eff}}(ps) &= -\frac{1}{(2\pi)^3\varepsilon_0}\int \mathrm{d}^3k \\
&\times \left[\frac{\displaystyle\sum_{\lambda=1}^{2}\sum_{i=1}^{3}\sum_{j=1}^{3} p_{\mathrm{P}i}\big[\boldsymbol{e}_i\cdot\boldsymbol{e}_\lambda(\boldsymbol{k})\big] p_{\mathrm{S}j}\big[\boldsymbol{e}_j\cdot\boldsymbol{e}_\lambda(\boldsymbol{k})\big]\mathrm{e}^{\mathrm{i}\boldsymbol{k}\cdot(\boldsymbol{r}_\mathrm{P}-\boldsymbol{r}_\mathrm{S})}}{\left(\dfrac{\hbar k^2}{2m_\mathrm{P}}+\Omega\right)-\dfrac{3\hbar}{2m_{\mathrm{eS}}}\left(\dfrac{\pi}{a_\mathrm{S}}\right)^2} \right. \\
&\left. + \frac{\displaystyle\sum_{\lambda=1}^{2}\sum_{i=1}^{3}\sum_{j=1}^{3} p_{\mathrm{S}j}\big[\boldsymbol{e}_j\cdot\boldsymbol{e}_\lambda(\boldsymbol{k})\big] p_{\mathrm{P}i}\big[\boldsymbol{e}_i\cdot\boldsymbol{e}_\lambda(\boldsymbol{k})\big]\mathrm{e}^{\mathrm{i}\boldsymbol{k}\cdot(\boldsymbol{r}_\mathrm{S}-\boldsymbol{r}_\mathrm{P})}}{\left(\dfrac{\hbar k^2}{2m_\mathrm{P}}+\Omega\right)-\dfrac{3\hbar}{2m_{\mathrm{eP}}}\left(\dfrac{\pi}{a_\mathrm{P}}\right)^2} \right] \\
&= -\frac{1}{(2\pi)^3\varepsilon_0}\sum_{\lambda=1}^{2}\sum_{i=1}^{3}\sum_{j=1}^{3} p_{\mathrm{P}i}p_{\mathrm{S}j}\times\int \mathrm{d}^3k\big[\boldsymbol{e}_i\cdot\boldsymbol{e}_\lambda(\boldsymbol{k})\big]\big[\boldsymbol{e}_j\cdot\boldsymbol{e}_\lambda(\boldsymbol{k})\big] \\
&\quad \mathrm{e}^{\mathrm{i}\boldsymbol{k}\cdot\boldsymbol{r}}\left\{\frac{1}{\dfrac{\hbar}{2m_\mathrm{P}}\left[k^2-\left(\dfrac{3\pi^2 m_\mathrm{P}}{m_{\mathrm{eS}}a_\mathrm{S}^2}-\dfrac{2m_\mathrm{P}\Omega}{\hbar}\right)\right]} \right. \\
&\quad \left. + \frac{1}{\dfrac{\hbar}{2m_\mathrm{P}}\left[k^2+\left(\dfrac{3\pi^2 m_\mathrm{P}}{m_{\mathrm{eP}}a_\mathrm{P}^2}-\dfrac{2m_\mathrm{P}\Omega}{\hbar}\right)\right]}\right\} . \qquad \text{(D.87)}
\end{aligned}
$$

Define μ_l and μ_h by

$$
\mu_\mathrm{l}=\left(\frac{3\pi^2 m_\mathrm{P}}{m_{\mathrm{eS}}a_\mathrm{S}^2}-\frac{2m_\mathrm{P}\Omega}{\hbar}\right)^{1/2}, \quad \mu_\mathrm{h}=\left(\frac{3\pi^2 m_\mathrm{P}}{m_{\mathrm{eP}}a_\mathrm{P}^2}+\frac{2m_\mathrm{P}\Omega}{\hbar}\right)^{1/2} . \qquad \text{(D.88)}
$$

Since this equation shows that $\mu_\mathrm{l} < \mu_\mathrm{h}$, the suffixes l and h are used, which are the first letters of 'light' and 'heavy', respectively. We then use the formula

$$
\frac{1}{(2\pi)^3}\int\frac{\exp(\mathrm{i}\boldsymbol{k}\cdot\boldsymbol{r})}{k^2+m^2}\mathrm{d}^3k=\frac{\exp(-mr)}{4\pi r} ,
$$

which is derived as follows. Cartesian coordinates (k_x, k_y, k_z) are fixed by taking the direction of k_z along that of $\boldsymbol{r}$, as shown in Fig. D.1. Transforming this to spherical coordinates, one derives relations $\mathrm{d}^3k = k^2\sin\theta\mathrm{d}k\mathrm{d}\theta\mathrm{d}\phi$ and $\boldsymbol{k} = (k\sin\theta\cos\phi, k\sin\theta\sin\phi, k\cos\theta)$. The right-hand side of the formula becomes

$$\frac{1}{(2\pi)^3}\int\frac{\exp(\mathrm{i}\boldsymbol{k}\cdot\boldsymbol{r})}{k^2+m^2}\mathrm{d}^3k = \frac{1}{(2\pi)^3}\int_0^{2\pi}\mathrm{d}\phi\int_0^{\pi}\sin\theta\mathrm{d}\theta\int_0^{\infty}\mathrm{d}k\frac{k^2\exp(ikr\cos\theta)}{k^2+m^2}$$
$$= \frac{1}{(2\pi)^2}\int_0^{\infty}\mathrm{d}k\int_0^{\pi}\sin\theta\mathrm{d}\theta\mathrm{e}^{ikr\cos\theta}\frac{k^2}{k^2+m^2}$$

by using these relations. Carrying out the integral with respect to θ, one derives

$$\frac{1}{(2\pi)^2}\int_0^{\infty}\mathrm{d}k\int_0^{\pi}\sin\theta\mathrm{d}\theta\mathrm{e}^{ikr\cos\theta}\frac{k^2}{k^2+m^2}$$
$$= \frac{1}{(2\pi)^2}\int_0^{\infty}\mathrm{d}k\frac{k^2}{k^2+m^2}\left[\frac{\mathrm{e}^{ikr}-\mathrm{e}^{-ikr}}{ikr}\right]$$
$$= \frac{1}{(2\pi)^2\mathrm{i}r}\left[\int_0^{\infty}\mathrm{d}k\frac{k\mathrm{e}^{ikr}}{k^2+m^2}+\int_{-\infty}^{0}\mathrm{d}k\frac{k\mathrm{e}^{ikr}}{k^2+m^2}\right]$$
$$= \frac{1}{(2\pi)^2\mathrm{i}r}\int_{-\infty}^{\infty}\mathrm{d}k\frac{k\mathrm{e}^{ikr}}{k^2+m^2}\,.$$

The integral with respect to k in the last line is reduced to obtain residues of the complex integral. Noting that the residue at $k = im$ is $\mathrm{e}^{-mr}/2$, one derives

$$\int_{-\infty}^{\infty}\mathrm{d}k\frac{k\mathrm{e}^{ikr}}{k^2+m^2} = 2\pi\mathrm{i}\frac{\mathrm{e}^{-mr}}{2} = \pi\mathrm{i}\mathrm{e}^{-mr}\,.$$

Substituting it into the equation given above, one has the required integration formula. Similarly, one can also derive the relation

$$\frac{1}{(2\pi)^3}\int\frac{\exp(\mathrm{i}\boldsymbol{k}\cdot\boldsymbol{r})}{k^2-m^2}\mathrm{d}^3k = \frac{\exp(imr)}{4\pi r}\,.$$

It follows that

$$\frac{1}{(2\pi)^3}\int\mathrm{d}^3k\mathrm{e}^{\mathrm{i}\boldsymbol{k}\cdot\boldsymbol{r}}\left(\frac{1}{k^2-\mu_\mathrm{l}^2}+\frac{1}{k^2+\mu_\mathrm{h}^2}\right) = \frac{\exp(\mathrm{i}\mu_\mathrm{l}r)}{4\pi r}+\frac{\exp(-\mu_\mathrm{h}r)}{4\pi r}\,. \quad \text{(D.89)}$$

Thus, neglecting the second term of the following equation

$$\sum_{\lambda=1}^{2}\left[\boldsymbol{e}_i\cdot\boldsymbol{e}_\lambda(\boldsymbol{k})\right]\left[\boldsymbol{e}_j\cdot\boldsymbol{e}_\lambda(\boldsymbol{k})\right] = \delta_{ij}(\boldsymbol{e}_i\cdot\boldsymbol{e}_j)-\left(\boldsymbol{e}_i\cdot\frac{\boldsymbol{k}}{k}\right)\left(\boldsymbol{e}_j\cdot\frac{\boldsymbol{k}}{k}\right)$$
$$= \boldsymbol{e}_i\cdot\boldsymbol{e}_i+\frac{(\boldsymbol{e}_i\cdot\nabla)(\boldsymbol{e}_i\cdot\nabla)}{k^2}\,,$$

$V_\mathrm{eff}(ps)$ is given by

$$V_\mathrm{eff}(ps) = -\frac{1}{4\pi\varepsilon_0}\sum_{i,j=1}^{3}p_{\mathrm{P}i}p_{\mathrm{S}j}\delta_{ij}\left[\frac{\exp(-\mu_\mathrm{h}r)}{r}+\frac{\exp(\mathrm{i}\mu_\mathrm{l}r)}{r}\right]$$
$$\propto \frac{\exp(-\mu_\mathrm{h}r)}{r}+\frac{\exp(\mathrm{i}\mu_\mathrm{l}r)}{r}\,. \quad \text{(D.90)}$$

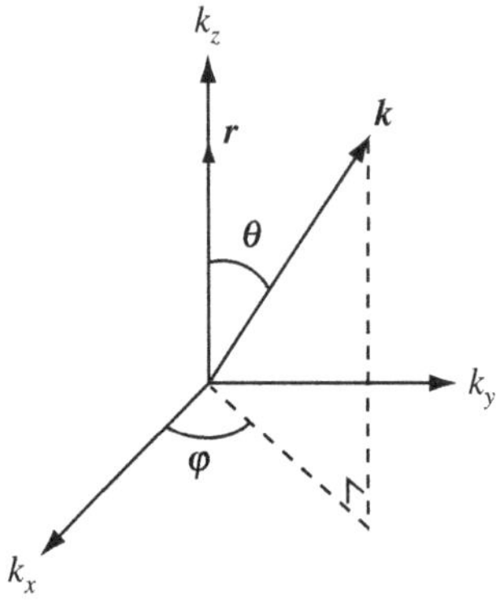

Fig. D.1. Spherical coordinate system in the $\boldsymbol{k}$-space

Since the eigenenergies of the probe and sample [$\hbar\Omega_0(p)$ and $\hbar\Omega_0(s)$ of (8.10a) and (8.10b), respectively] are larger than that of the macroscopic matter $\hbar\Omega$, μ_l and μ_h of (D.88) are approximated as

$$\mu_\mathrm{l} \approx \sqrt{\frac{3\pi^2 m_\mathrm{P}}{m_\mathrm{eS} a_\mathrm{S}^2}}\,, \quad \mu_\mathrm{h} \approx \sqrt{\frac{3\pi^2 m_\mathrm{P}}{m_\mathrm{eP} a_\mathrm{P}^2}}\,. \tag{D.91}$$

Finally, it follows that

$$V_\mathrm{eff}(ps) \propto \frac{\exp(-\mu_\mathrm{P}\pi r/a_\mathrm{P})}{r} + \frac{\exp(\mathrm{i}\mu_\mathrm{S}\pi r/a_\mathrm{S})}{r}\,, \tag{D.92}$$

which is none other than (8.12).

Solutions to Problems

Chapter 1

Problem 1.1

(a) The magnitude of the electric field of a spherical light wave propagating from the point P to Q is expressed as $u_0[\exp(\mathrm{i}kr)/r]\cos\psi$, where u_0 is a constant proportional to the amplitude and k is the wave number ($\equiv 2\pi/\lambda$). Since point P can be located everywhere in the slit area, the total amplitude $u(x_2, z)$ of the cylindrical light wave diverging from the slit is given by integrating $u_0[\exp(\mathrm{i}kr)/r]\cos\psi$ over the slit, i.e.,

$$u(x_2, z) = \int_{-a/2}^{a/2} u_0 \frac{\exp(\mathrm{i}kr)}{r} \cos\psi \mathrm{d}x_1 \ . \tag{Q1.1}$$

Several approximations have to be made in order to carry out this integration. First, $\cos\psi$ is approximated as unity because $\psi \approx 0$. Second, r in the denominator of $\exp(\mathrm{i}kr)/r$ can be approximated as z because $\psi \approx 0$. However, kr in the numerator must not be approximated as kz because $r \gg \lambda$. For kr, a more accurate approximation must be made by expanding r as

$$\begin{aligned} r &= z\sqrt{1 + \left(\frac{x_2 - x_1}{z}\right)^2} \\ &= z\left[1 + \frac{1}{2}\left(\frac{x_2 - x_1}{z}\right)^2 + \cdots\right] \\ &= z + \frac{x_2^2}{2z} - \frac{x_2 x_1}{z} + \frac{x_1^2}{2z} + \cdots \ . \end{aligned} \tag{Q1.2}$$

The formula $\sqrt{1+x} = 1 + x/2 + \cdots$ has been used to transform from the first to the second line.

Using the first three terms in (Q1.2) to approximate kr, the final form of (Q1.1) is

$$u(x_2, z) = A_0(x_2, z) \int_{-a/2}^{a/2} u_0 \exp\left(-\mathrm{i}\frac{kx_2}{z}x_1\right) \mathrm{d}x_1 \ , \tag{Q1.3}$$

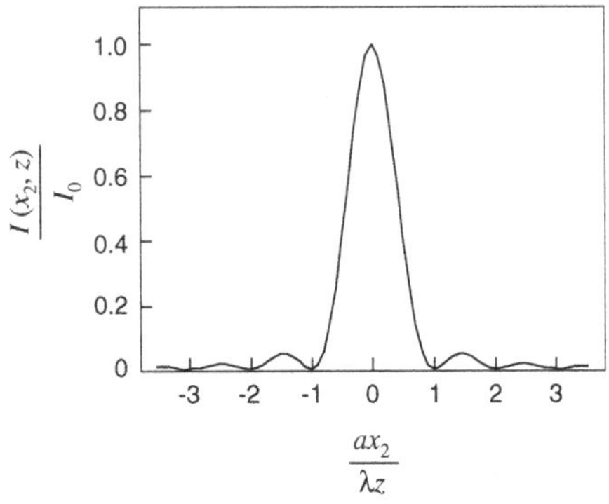

Fig. Q1.1. Relation between $ax_2/\lambda z$ and $I(x_2, z)$ given by (Q1.7)

where $A_0(x_2, z)$ is a quantity independent of x_1. The upper and lower limits of the integral in (Q1.3) can be extended to $+\infty$, and $-\infty$, respectively because u_0 is zero outside the slit. Furthermore, defining the variable f_x by $f_x \equiv x_2/\lambda z$, (Q1.3) is transformed to

$$u(x_2, z) = A_0(x_2, z) \int_{-\infty}^{\infty} u_0 \exp(-i2\pi f_x x_1) dx_1 \, . \tag{Q1.4}$$

Note that this integral represents the Fourier transform from the space x_1 to the space f_x, where f_x is called a spatial Fourier frequency. This means that the spatial distribution of amplitude $u(x_2, z)$ on $\sum'$ is the Fourier transform of the amplitude u_0 on $\sum$.

Performing the integration in (Q1.3) gives

$$u(x_2, z) = A_0(x_2, z) u_0 \operatorname{sinc}\left(\frac{ax_2}{\lambda z}\right) , \tag{Q1.5}$$

where

$$\operatorname{sinc} X \equiv \frac{\sin \pi X}{\pi X} \, . \tag{Q1.6}$$

The diffraction pattern, i.e., the spatial distribution of the light intensity $I(x_2, z)$ to be observed on $\sum'$ is proportional to $|u(x_2, z)|^2$, which is derived from (Q1.5) and is expressed as

$$I(x_2, z) = I_0 \operatorname{sinc}^2\left(\frac{ax_2}{\lambda z}\right) , \tag{Q1.7}$$

where $I_0 \equiv |A_0(x_2, z) u_0|^2$. This distribution represents the diffraction pattern, and gives the solution to part (a) of the problem. Figure Q1.1 shows the dependence of $I(x_2, z)$ on x_2.

(b) The variable x_2 representing the point Q on $\sum'$ is expressed as $x_2 = z \tan\theta$ in polar coordinates, where the center of the slit on $\sum$ is taken as the origin for these coordinates. Using a paraxial approximation, i.e., $\theta \approx 0$ and $\tan\theta \approx \theta$, the horizontal axis in Fig. Q1.1 is transformed from x_2 to θ using the relation

$$\frac{ax_2}{\lambda z} \approx \frac{a\theta}{\lambda} . \qquad \text{(Q1.8)}$$

After this transformation, the full width at half maximum $\Delta\theta_{\mathrm{h}}$ of the light intensity in Fig. Q1.1 is given by

$$\Delta\theta_{\mathrm{h}} = \frac{0.89\lambda}{a} . \qquad \text{(Q1.9)}$$

This is the solution to problem (b) because $\Delta\theta_{\mathrm{h}}$ can be considered as the divergence angle.

This equation shows that the divergence angle is proportional to λ/a, which implies that the diffraction effect is more prominent for longer wavelengths and smaller slit sizes. This is valid not only for a slit but also for rectangular, circular or other shapes of aperture.

Problem 1.2

When light is focused by a lens, the spot size suffers from the effects of diffraction because the lens diameter is finite. By considering a of (Q1.9) as the width of the cylindrical lens, the full width at half maximum of the diffraction pattern on the focal plane is expressed as

$$\Delta x_{\mathrm{h}} \approx f\Delta\theta_{\mathrm{h}} = \frac{0.89\lambda}{a} = \frac{0.45\lambda}{\sin\alpha} , \qquad \text{(Q1.10)}$$

where f is the focal length and $\sin\alpha$ is defined by $(a/2)/\sqrt{f^2 + (a/2)^2}$. In terms of the numerical aperture NA, Δx_{h} of (Q1.10) is given approximately by λ/NA. This representation is valid not only for the cylindrical lens but also for a circular convex lens.

Chapter 2

Problem 2.1

The answer is 'no'. Even though the fiber probe has the profile of Fig. 2.11, it has a finer structure which is observed if the profile is magnified as shown in Fig. Q2.1a. Thus, when using this fiber probe in illumination mode, it generates various optical near fields whose spatial distributions depend on the size of the finer structure. Therefore, bringing the fiber probe to within a distance equivalent to this size from the sample, a higher resolution is obtained, indeed, much higher than that determined by a_{f}. Figures Q2.1b and 2.8 show the protruded-type fiber probe devised as a result of this discussion.

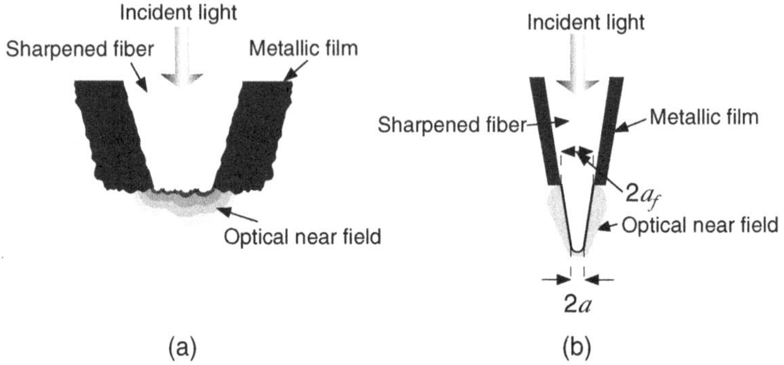

Fig. Q2.1. Schematic diagrams used to discuss the characteristics of the fiber probe in Fig. 2.11. (**a**) Magnified drawing of the top of the fiber probe in Fig. 2.11. (**b**) Profile of a fiber probe designed to realize higher resolution

Since the sharpened core with small tip radius a protrudes from the metallic film, one can obtain high resolution determined by a. The metallic film on the foot of the fiber probe is used to screen the scattered light 1, whose foot radius a_f does not determine the resolution.

Figure 2.11 illustrates the output light as flowing out from the fiber probe in a way analogous to tap water. However, it corresponds to the scattered light 1. The optical near field must be illustrated as a bubble hanging down from a straw, as shown in Figs. Q2.1a and b.

Problem 2.2

The wavelength of light is not directly correlated to color. The color is correlated to the photon energy, which is proportional to the frequency of light. The optical near field and scattered light 2 shown in Fig. 2.6b have the same frequency as that of the incident light as long as resonance interaction, photon emission, and changes in electronic energy states are not involved in the electromagnetic interaction between spheres S and P. Thus, the color of the optical near field is the same as that of the incident light.

Chapter 3

Problem 3.1

The answer is 'no'. This is because the process of optical-near-field detection involves light scattering between two closely spaced particles, as shown in Fig. 2.6b. Since the diffraction grating has a periodic structure, the light scattering becomes a many-body phenomenon because adjacent corrugations are involved. Thus, evaluation of resolution becomes complicated.

The conventional optical microscope has low-pass filtering characteristics due to the diffraction limit, as shown in Fig. 3.2b, while the near-field optical microscope has band-pass characteristics. Thus, evaluation of the resolution of a near-field optical microscope requires a novel method which differs from the one used for conventional optical microscopes. It is most suitable to define the resolution as the reciprocal of the high frequency cutoff of the spatial power spectral density calculated from the Fourier transform of the image profile. For such calculations, the optimum sample is the one corresponding to the spatially white noise, i.e., mutually isolated randomly-sized particles fixed at random positions on a substrate.

Since the diffraction grating has a specific spatial Fourier frequency component, its spatial power spectral density corresponds to the line spectrum, and so is not suitable for evaluating the resolution. Further, since the resolution and efficiency of optical-near-field detection depends on the characteristics of the probe, the standard sample for evaluating the resolution should be determined as a function of the probe. A universal standard sample does not exist for the near-field optical microscope.

Problem 3.2

The answer is 'no'. High resolution cannot always be obtained. When measuring photoluminescence from a semiconductor, photons are emitted after the carriers diffuse for about 1 μm, even though the semiconductor surface is locally excited by the optical near field. Thus, the image size is as large as the diffusion length if the emitted photons are collected by a convex lens. The emitted photons must be collected through the probe in order to avoid the effects of carrier diffusion on the resolution. By this detection, resolution as high as the size of the probe tip is obtained. This is the reason why the experimental setup in Fig. 3.10 combines the illumination and collection modes.

Chapter 4

Problem 4.1

Figure Q4.1 shows that the two particles with charges $+q$ and $-q$ are fixed with separation d. Their positions are expressed as $(0, 0, d/2)$ and $(0, 0, -d/2)$, respectively. We calculate here the electric field at the position $A(x, y, z)$ generated by these particles, where $r = \sqrt{x^2 + y^2 + z^2} \gg d$. The vectors r_1 and r_2 oriented from these particles to the position A are given by

$$\boldsymbol{r}_1 = \left(x, y, z - \frac{d}{2}\right) \quad \text{and} \quad \boldsymbol{r}_2 = \left(x, y, z + \frac{d}{2}\right) . \tag{Q4.1}$$

Using these vectors, the electric field derived from Coulomb's law is

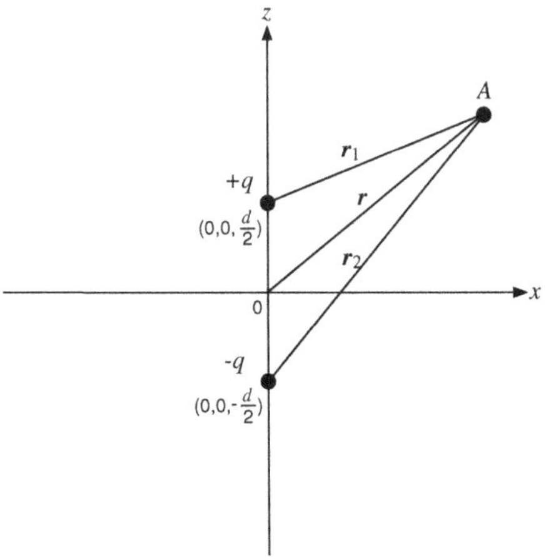

Fig. Q4.1. Configuration of the two particles with charges $+q$ and $-q$

$$\boldsymbol{E}(\boldsymbol{r}) = \frac{q}{4\pi\varepsilon_0}\left(\frac{\boldsymbol{r}_1}{r_1^3} - \frac{\boldsymbol{r}_2}{r_2^3}\right) , \tag{Q4.2}$$

where $r_1 = |\boldsymbol{r}_1|$ and $r_2 = |\boldsymbol{r}_2|$. Noting that $r \gg d$, r_1 and r_2 can be approximated by

$$r_1^{-3} \approx r^{-3}\left(1 + \frac{3}{2}\frac{zd}{r^2}\right) \quad \text{and} \quad r_2^{-3} \approx r^{-3}\left(1 - \frac{3}{2}\frac{zd}{r^2}\right) . \tag{Q4.3}$$

Substituting (Q4.1) and (Q4.3) into (4.2), the z-component of the electric field $\boldsymbol{E}(\boldsymbol{r})$ is derived as

$$E_z(\boldsymbol{r}) = \frac{1}{4\pi\varepsilon_0}\frac{qd}{r^3}\left(\frac{3z^2}{r^2} - 1\right) . \tag{Q4.4}$$

The x-component can also be derived as

$$E_x(\boldsymbol{r}) = \frac{1}{4\pi\varepsilon_0}\frac{qd}{r^3}\frac{3xz}{r^2} . \tag{Q4.5}$$

The y-component $E_y(\boldsymbol{r})$ is derived if x in (Q4.5) is replaced by y.

The above result shows that the magnitude of the electric field is proportional to the product of the charge and the separation qd. Thus, using a unit vector $\boldsymbol{n} = \boldsymbol{r}/|\boldsymbol{r}|$ and the electric dipole moment defined by

$$\boldsymbol{p} = (0, 0, qd) , \tag{Q4.6}$$

equations (Q4.4) and (Q4.5) give

$$\boldsymbol{E}(\boldsymbol{r}) = \frac{1}{4\pi\varepsilon_0}\left[3(\boldsymbol{p}\cdot\boldsymbol{n})\boldsymbol{n} - \boldsymbol{p}\right]\frac{1}{r^3} . \tag{Q4.7}$$

This is equivalent to (A.28) with $k = 0$.

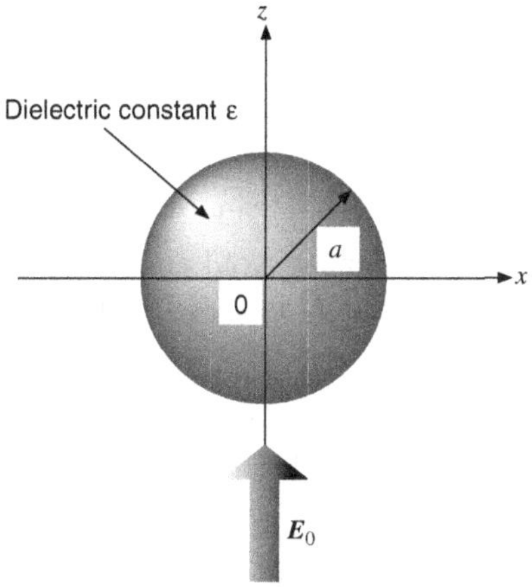

Fig. Q4.2. Sphere with radius a and dielectric constant ε illuminated by a homogeneous electric field $\boldsymbol{E}_0$ in vacuum

Problem 4.2

Using the same symbols as in Problem 4.1, the sum of the electric potential at position A generated by the two charged particles is given by

$$V = \frac{1}{4\pi\varepsilon_0}\left(\frac{q}{r_1^3} - \frac{q}{r_2^3}\right) . \tag{Q4.8}$$

Using (Q4.3), this approximates to

$$V \approx \frac{1}{4\pi\varepsilon_0}\frac{qd}{r^2}\frac{z}{r} . \tag{Q4.9}$$

Using the electric dipole moment $\boldsymbol{p}$ of (Q4.6) and the unit vector $\boldsymbol{n}$, it is transformed to

$$V = \frac{1}{4\pi\varepsilon_0}\frac{\boldsymbol{p}\cdot\boldsymbol{n}}{r^2} . \tag{Q4.10}$$

Problem 4.3

Figure Q4.2 shows a sphere with radius a and dielectric constant ε illuminated by a homogeneous electric field $\boldsymbol{E}_0$ in vacuum. In spherical coordinates (r, θ, ϕ: $x = r\sin\theta\cos\phi$, $y = r\sin\theta\sin\phi$, $z = r\cos\theta$), the solution of Laplace's equation [$\rho = 0$ in (A.4)] gives the electric potential outside the sphere, viz.,

$$V_{\mathrm{out}} = Ar\cos\theta + \frac{B}{r^2}\cos\theta . \tag{Q4.11}$$

The potential inside the sphere is given by

$$V_{\text{in}} = Cr\cos\theta + \frac{D}{r^2}\cos\theta\ . \tag{Q4.12}$$

Since the electric potential V_{out} is independent of the sphere at $r \to \infty$, the constant A in (Q4.11) is

$$A = -E_0\ , \tag{Q4.13}$$

where $E_0 \equiv |\boldsymbol{E}_0|$. Then, in order to avoid the value of V_{in} diverging at the center of the sphere, the constant D in (Q4.12) must be

$$D = 0\ . \tag{Q4.14}$$

Further, since the electric potential is continuous ($V_{\text{in}} = V_{\text{out}}$) at the surface of the sphere ($r = a$), the relation

$$Aa + \frac{B}{a^2} = Ca \tag{Q4.15}$$

is valid. On the other hand, the normal component of the electric flux density D is also continuous at $r = a$ and can be expressed as

$$-\varepsilon_0 \left.\frac{\partial V_{\text{out}}}{\partial r}\right|_{r=a} = -\varepsilon \left.\frac{\partial V_{\text{in}}}{\partial r}\right|_{r=a} \tag{Q4.16}$$

using the relation $\boldsymbol{D} = \varepsilon\boldsymbol{E} = -\varepsilon\nabla V$, where ∇ is as defined in Sect. A.1. Substituting (Q4.11) and (Q4.12) into this equation, one obtains

$$\varepsilon_0 A + \frac{2\varepsilon_0}{a^3}B = -\varepsilon C\ . \tag{Q4.17}$$

The values of B and C can be obtained by solving (Q4.13), (Q4.15), and (Q4.17) simultaneously. They are

$$B = \frac{\varepsilon - \varepsilon_0}{\varepsilon + 2\varepsilon_0}a^3 E_0\ , \tag{Q4.18a}$$

$$C = -\frac{3\varepsilon_0}{\varepsilon + 2\varepsilon_0}E_0\ . \tag{Q4.18b}$$

In summary, the values of V_{in}, E_{in}, and D_{in} in the sphere are

$$V_{\text{in}} = -\frac{3\varepsilon_0}{\varepsilon + 2\varepsilon_0}E_0 r\cos\theta\ , \tag{Q4.19}$$

$$\boldsymbol{E}_{\text{in}} = \frac{3\varepsilon_0}{\varepsilon + 2\varepsilon_0}\boldsymbol{E}_0\ , \tag{Q4.20}$$

and

$$\boldsymbol{D}_{\text{in}} = \varepsilon\boldsymbol{E}_{\text{in}} = \frac{3\varepsilon}{\varepsilon + 2\varepsilon_0}\boldsymbol{D}_0\ . \tag{Q4.21}$$

They indicate that $|\boldsymbol{E}_{\text{in}}| < |\boldsymbol{E}_{\text{out}}|$ and $|\boldsymbol{D}_{\text{in}}| > |\boldsymbol{D}_{\text{out}}|$ for $\varepsilon < \varepsilon_0$. On the other hand, the value of V_{out} outside the sphere is

$$V_{\text{out}} = -E_0 r \cos\theta + \frac{\varepsilon - \varepsilon_0}{\varepsilon + 2\varepsilon_0} E_0 \frac{a^3}{r^2} \cos\theta \,. \tag{Q4.22}$$

It is equal to the sum of the electric potential due to the applied electric field $\boldsymbol{E}_0$ and that of the electric dipole moment

$$\boldsymbol{p} = 4\pi\varepsilon_0 \left(\frac{\varepsilon - \varepsilon_0}{\varepsilon + 2\varepsilon_0} \right) a^3 \boldsymbol{E}_0 \,, \tag{Q4.23}$$

at the center of the sphere, which is given by

$$V_{\text{out}} = -E_0 r \cos\theta + \frac{1}{4\pi\varepsilon_0} \frac{p}{r^2} \cos\theta \,. \tag{Q4.24}$$

(The electric potential due to an electric dipole moment is given in Problem 4.2.) Equation (Q4.23) indicates that the polarizability α of the sphere is

$$\alpha = 4\pi\varepsilon_0 \left(\frac{\varepsilon - \varepsilon_0}{\varepsilon + 2\varepsilon_0} \right) a^3 \,. \tag{Q4.25}$$

Problem 4.4

Since Sect. 4.2.2 shows that the value of the half width at half maximum x_{h} is $0.77(a_{\text{P}} + a_{\text{S}} + z)$ when scanning the sphere P at elevation z, the relation $z \leq a_{\text{P}} + a_{\text{S}}$ must be valid in order to obtain a value of x_{h} less than twice that of (4.15). This means that the maximum allowable value of z is $a_{\text{P}} + a_{\text{S}}$.

Chapter 5

Problem 5.1

The electric potential V_{vac} in vacuum can be derived by summing the contributions from the physical charge $+q$ at A and the fictitious charge $-q'$ at B, which gives

$$V_{\text{vac}} = \frac{1}{4\pi\varepsilon_0} \left[\frac{q}{\sqrt{(x-a)^2 + y^2 + z^2}} - \frac{q'}{\sqrt{(x+a)^2 + y^2 + z^2}} \right] \,. \tag{Q5.1}$$

Due to the contribution from the fictitious charge $+q''$ at A, the electric potential V_{diel} in the dielectric is given by

$$V_{\text{diel}} = \frac{1}{4\pi\varepsilon_0} \frac{q''}{\sqrt{(x-a)^2 + y^2 + z^2}} \,. \tag{Q5.2}$$

By the continuity of the electric potential at the boundary between the vacuum and dielectric, i.e., $V_{\text{vac}}(x=0) = V_{\text{diel}}(x=0)$, equations (Q5.1) and (Q5.2) give

$$q - q' = q'' . \tag{Q5.3}$$

In order to represent the continuity of the normal component (the x-component) D_x of the electric flux density at the boundary, (Q5.1) is substituted into the relation $E_x = -\partial V/\partial x$ [refer to (A.1) of Appendix A] at $x = 0$. As a result, one obtains

$$\begin{aligned}(D_x)_{\text{vac}} &= \varepsilon_0 (E_x)_{\text{vac}} \\ &= -\frac{1}{4\pi}\left[\frac{aq}{(a^2+y^2+z^2)^{3/2}} + \frac{aq'}{(a^2+y^2+z^2)^{3/2}}\right] .\end{aligned} \tag{Q5.4}$$

Similarly, substituting (Q5.2) at $x = 0$, one obtains

$$(D_x)_{\text{diel}} = \varepsilon (E_x)_{\text{diel}} = -\frac{\varepsilon}{4\pi\varepsilon_0}\frac{aq''}{(a^2+y^2+z^2)^{3/2}} . \tag{Q5.5}$$

By these equations, continuity is expressed as

$$q + q' = \frac{\varepsilon}{\varepsilon_0} q'' . \tag{Q5.6}$$

Finally from (Q5.3) and (Q5.6), one obtains

$$q' = \frac{\varepsilon - \varepsilon_0}{\varepsilon + \varepsilon_0} q , \tag{Q5.7a}$$

$$q'' = \frac{2\varepsilon_0}{\varepsilon + \varepsilon_0} q . \tag{Q5.7b}$$

In summary, fictitious charges $-q'$ and $+q''$ can be used to derive the electric field in vacuum and in the dielectric, respectively, and the boundary condition can then be removed.

Chapter 6

Problem 6.1

The right-hand side of (6.12) can be obtained by substituting $\boldsymbol{T}(\boldsymbol{r}, \boldsymbol{r}')$ of (6.10) into the left-hand side of (6.12). Here, note that $G(\boldsymbol{r}, \boldsymbol{r}')$ of (6.5) satisfies (6.6), and use the formula $\nabla \times \nabla \times \boldsymbol{T} = \nabla(\nabla \cdot \boldsymbol{T}) - \nabla^2 \boldsymbol{T}$.

Problem 6.2

Note that the differential operator ∇_R is expressed as $(\boldsymbol{R}/R)(\mathrm{d}/\mathrm{d}R)$. Then (6.13) is derived by substituting (6.5) into (6.10). Here, use the relation $\nabla_R\left[\exp(\mathrm{i}kR)/R\right] = \left\{\mathrm{i}k\left[\exp(\mathrm{i}kR)/R\right] - \left[\exp(\mathrm{i}kR)/R^2\right]\right\}(\boldsymbol{R}/R)$, and calculate $\nabla_R\left\{\nabla_R\left[\exp(\mathrm{i}kR)/R\right]\right\}$.

Problem 6.3

The electric field of (A.28) is given by $\boldsymbol{T}(\boldsymbol{r},\boldsymbol{r}')\boldsymbol{p}(\boldsymbol{r}')$. Further, note that $\boldsymbol{r}'=0$ and $\boldsymbol{R}=\boldsymbol{r}$ because the electric dipole in (A.28) is fixed at the origin. Then, abbreviating $\boldsymbol{p}(0)$ to $\boldsymbol{p}$, one can derive the electric field from (6.13) as

$$\begin{aligned}\boldsymbol{E}(\boldsymbol{r}) &= \boldsymbol{T}(\boldsymbol{r},0)\boldsymbol{p} \\ &= \frac{1}{4\pi\varepsilon_0}\left[k^2[\boldsymbol{p}-\boldsymbol{n}(\boldsymbol{n}\cdot\boldsymbol{p})]\frac{1}{r}+[3\boldsymbol{n}(\boldsymbol{n}\cdot\boldsymbol{p})-\boldsymbol{p}]\left(-\frac{\mathrm{i}k}{r^2}+\frac{1}{r^3}\right)\right]\mathrm{e}^{\mathrm{i}kr} \\ &= \frac{1}{4\pi\varepsilon_0}\left[k^2(\boldsymbol{n}\times\boldsymbol{p})\times\boldsymbol{n}\left(\frac{1}{r}\right)+[3\boldsymbol{n}(\boldsymbol{n}\cdot\boldsymbol{p})-\boldsymbol{p}]\left(-\frac{\mathrm{i}k}{r^2}+\frac{1}{r^3}\right)\right]\mathrm{e}^{\mathrm{i}kr}\,,\end{aligned} \tag{Q6.1}$$

which is equal to (A.28). The formula $(\boldsymbol{n}\times\boldsymbol{p})\times\boldsymbol{n}=(\boldsymbol{n}\cdot\boldsymbol{n})\boldsymbol{p}-(\boldsymbol{n}\cdot\boldsymbol{p})\boldsymbol{n}$ was used to derive the last term on the right-hand side of this equation.

Problem 6.4

The single electric dipole moment $\boldsymbol{p}(\boldsymbol{r}')$ at the position $\boldsymbol{r}'$ is taken as the simplest source of electromagnetic fields and the propagator $\boldsymbol{T}(\boldsymbol{r},\boldsymbol{r}')$ is applied to it. Then the generated electric field $\boldsymbol{E}(\boldsymbol{r})$ is expressed by (6.11) as $\boldsymbol{T}(\boldsymbol{r},\boldsymbol{r}')\boldsymbol{p}(\boldsymbol{r}')$. Using this expression and (6.13), the x-component of the electric field is

$$\begin{aligned}E_x &= \frac{1}{4\pi\varepsilon_0}\left[k^2(p_x-n_x\boldsymbol{n}\cdot\boldsymbol{p})\frac{1}{R}+(3n_x\boldsymbol{n}\cdot\boldsymbol{p}-p_x)\left(-\frac{\mathrm{i}k}{R^2}+\frac{1}{R^3}\right)\right]\mathrm{e}^{\mathrm{i}kR} \\ &= \frac{1}{4\pi\varepsilon_0}\left[k^2[p_x-n_x(n_xp_x+n_yp_y+n_zp_z)]\frac{1}{R}\right. \\ &\quad \left.+[3n_x(n_xp_x+n_yp_y+n_zp_z)-p_x]\left(-\frac{\mathrm{i}k}{R^2}+\frac{1}{R^3}\right)\right]\mathrm{e}^{\mathrm{i}kR}\,.\end{aligned} \tag{Q6.2}$$

Thus, comparing this equation with $E_x=T_{xx}p_x+T_{xy}p_y+T_{xz}p_z$, one derives

$$T_{xx} = \frac{1}{4\pi\varepsilon_0}\left[k^2(1-n_x^2)\frac{1}{R}+(3n_x^2-1)\left(-\frac{\mathrm{i}k}{R^2}+\frac{1}{R^3}\right)\right]\mathrm{e}^{\mathrm{i}kR}\,, \tag{Q6.3a}$$

$$T_{xy} = \frac{1}{4\pi\varepsilon_0}n_xn_y\left[-k^2\frac{1}{R}+3\left(-\frac{\mathrm{i}k}{R^2}+\frac{1}{R^3}\right)\right]\mathrm{e}^{\mathrm{i}kR}\,, \tag{Q6.3b}$$

$$T_{xz} = \frac{1}{4\pi\varepsilon_0}n_xn_z\left[-k^2\frac{1}{R}+3\left(-\frac{\mathrm{i}k}{R^2}+\frac{1}{R^3}\right)\right]\mathrm{e}^{\mathrm{i}kR}. \tag{Q6.3c}$$

Equation (6.15) can be obtained once other elements have been derived in a similar manner.

Chapter 7

Problem 7.1

Substitute $\Phi = [\varepsilon(\boldsymbol{r}) - \varepsilon_0]/\varepsilon_0$ and $\boldsymbol{A} = \nabla\phi(\boldsymbol{r})$ into the vector analysis formula $\nabla\cdot(\Phi\boldsymbol{A}) = \Phi\nabla\cdot\boldsymbol{A} + \boldsymbol{A}\cdot\nabla\Phi$.

Problem 7.2

Substitute $\Phi = 1 - \varepsilon_0/\varepsilon(\boldsymbol{r})$ and $\boldsymbol{A} = \nabla \times \boldsymbol{C}$ into the vector analysis formula $\nabla \times (\Phi\boldsymbol{A}) = \Phi\nabla \times \boldsymbol{A} + \nabla\Phi \times \boldsymbol{A}$.

Chapter 8

Problem 8.1

To carry out the integral in (8.16), the Cartesian coordinate system (x', y', z') is defined by taking the z'-axis along the vector r. Then transforming to a spherical coordinate system (r', θ', ϕ'), one has $\mathrm{d}^3r' = r'^2 \sin\theta' \mathrm{d}\theta' \cos\phi' \mathrm{d}\phi'$, $\boldsymbol{r} = (0, 0, r)$, and $\boldsymbol{r}' = (r'\sin\theta'\cos\phi', r'\sin\theta'\sin\phi', r'\cos\theta')$. Noting that $r = |\boldsymbol{r}| > a$ and $r' = |\boldsymbol{r}'| < a$, (8.16) becomes

$$
\begin{aligned}
&\int_{\text{sphere}} \mathrm{d}^3r' \frac{\exp(-\mu|\boldsymbol{r} - \boldsymbol{r}'|)}{|\boldsymbol{r} - \boldsymbol{r}'|} \\
&= \int_0^a r'^2 \mathrm{d}r' \int_0^\pi \sin\theta' \mathrm{d}\theta' \int_0^{2\pi} \mathrm{d}\phi' \frac{\exp\left[-\mu\sqrt{r'^2 \sin^2\theta' + (r - r'\cos\theta')^2}\right]}{\sqrt{r'^2 \sin^2\theta' + (r - r'\cos\theta')^2}} \\
&= 2\pi \int_0^a r'^2 \mathrm{d}r' \int_{r-r'}^{r+r'} \frac{\xi \mathrm{d}\xi}{rr'} \frac{\exp(-\mu\xi)}{\xi} \\
&= \frac{2\pi}{r} \int_0^a r' \mathrm{d}r' \int_{r-r'}^{r+r'} \mathrm{d}\xi \exp(-\mu\xi) ,
\end{aligned}
\tag{Q8.1}
$$

where the integration over θ' is transformed to an integration over ξ using relations $\xi^2 = r'^2 \sin^2\theta' + (r - r'\cos\theta')^2 = r^2 + r'^2 - 2rr'\cos\theta'$ and $\xi \mathrm{d}\xi = rr'\sin\theta' \mathrm{d}\theta'$. Further, carrying out the integration over ξ in the last term of (Q8.1), one derives

$$
\int_{r-r'}^{r+r'} \mathrm{d}\xi \mathrm{e}^{-\mu\xi} = \left[\frac{\exp(-\mu\xi)}{-\mu}\right]_{r-r'}^{r+r'} = \frac{1}{\mu}\left[\mathrm{e}^{-\mu(r-r')} - \mathrm{e}^{-\mu(r+r')}\right] . \tag{Q8.2}
$$

On the other hand, noting that

$$\int_0^a r'\mathrm{d}r'\mathrm{e}^{\pm\mu r'} = \left[\frac{r'\exp(\pm\mu r')}{\pm\mu}\right]_0^a - \int_0^a \mathrm{d}r'\frac{\exp(\pm\mu r')}{\pm\mu}$$
$$= \pm\frac{a}{\mu}\mathrm{e}^{\pm\mu a} \mp \left[\frac{\exp(\pm\mu r')}{\pm\mu}\right]_0^a$$
$$= \pm\frac{a}{\mu}\mathrm{e}^{\pm\mu a} - \frac{1}{\mu^2}(\mathrm{e}^{\pm\mu a} - 1)\ , \qquad \text{(Q8.3)}$$

substituting (Q8.2) and (Q8.3) into the last line of (Q8.1) leads to

$$\int_{\text{sphere}} \mathrm{d}^3r'\frac{\exp(-\mu|\boldsymbol{r}-\boldsymbol{r}'|)}{|\boldsymbol{r}-\boldsymbol{r}'|}$$
$$= \frac{2\pi}{\mu^3}\left\{(1+\mu a)\frac{\exp[-\mu(r+a)]}{r} - (1-\mu a)\frac{\exp[-\mu(r-a)]}{r}\right\}\ ,$$

which gives (8.18).

Chapter 9

Problem 9.1

The Rabi angular frequency $\Omega(r)$ in (9.2) depends on the distance r from the planar surface and is expressed as $\gamma\sqrt{I(r)/2I_\mathrm{s}}$. Since the optical-near-field intensity $I(r)$ depends on r and is expressed as $I_0\exp(-2r/\Lambda)$, $\Omega^2(r)$ is $\gamma^2 I_0\exp(-2r/\Lambda)/2I_\mathrm{s}$, implying $\mathrm{d}\Omega^2(r)/\mathrm{d}r = (-2/\Lambda)\Omega^2(r)$. Equation (9.2) follows if the gradient $\nabla\Omega^2(r)$ and the detuning δ of (9.1) are replaced by this relation and by the detuning Δ including the Doppler effect, respectively.

Problem 9.2

The optical potential $U_\mathrm{opt}(\rho,\phi)$ is given by the spatial integral of the dipole force $\boldsymbol{F}_\mathrm{d}$, which is expressed as

$$U_\mathrm{opt}(\rho,\phi) = \iiint(-\boldsymbol{F}_\mathrm{d})\mathrm{d}^3r\ . \qquad \text{(Q9.1)}$$

Noting that $\nabla\Omega^2$ of (9.1) is a spatial derivative of the square of the Rabi angular frequency Ω^2, $\nabla\Omega^2\mathrm{d}^3r$ is expressed as $\mathrm{d}\Omega^2$. Using this expression and (9.1), (Q9.1) becomes

$$U_\mathrm{opt}(\rho,\phi) = \int\frac{\hbar\Delta^2}{4\Delta^2+\gamma^2+2\Omega^2}\mathrm{d}\Omega^2\ , \qquad \text{(Q9.2)}$$

where the detuning δ is replaced by the detuning Δ including the Doppler effect. Setting $U_\mathrm{opt}(\rho,\phi) = 0$ at $\Omega = 0$, the integral of (Q9.2) gives

$$U_{\mathrm{opt}}(\rho, \phi) = \frac{\hbar \Delta}{2} \ln\left(1 + \frac{2\Omega^2}{4\Delta^2 + \gamma^2}\right) . \tag{Q9.3}$$

Equation (9.5) is derived by substituting the relation

$$\Omega^2(r) = \gamma^2 \frac{I_0}{2I_{\mathrm{s}}} \exp(-2r/\Lambda)$$

into this equation.

References

Chapter 1

1.1 I. Newton: *Optics: A Treatise of the Reflections, Refractions, Inflections and Colours of Light* (Royal Society, London, 1704)
1.2 C.H. Townes: *How the Laser Happened* (Oxford University Press, New York 1999)
1.3 M. Ohtsu: *Coherent Quantum Optics and Technology* (Kluwer Academic Publishers, Dordrecht/Boston/Tokyo 1992)
1.4 Optoelectronic Industry and Technology Development Association: *Report on Photonics Technology Road Map: Information Storage* (OITDA, Tokyo 1998)
1.5 Optoelectronic Industry and Technology Development Association: *Report on Photonics Technology Road Map: Information Communication* (OITDA, Tokyo 1998)
1.6 M. Ohtsu: *Modern Optical Science I* (Asakura-shoten, Tokyo 1994)

Chapter 3

3.1 E.A. Synge: Phil. Mag. **6**, 356 (1928)
3.2 M. Ohtsu: Overview in: *Near-Field Optics: Principles and Applications*, ed. by X. Zhu and M. Ohtsu (World Scientific, Singapore New Jersey 2000) pp. 1–8
3.3 D.W. Pohl, W. Denk, M. Lanz: Appl. Phys. Lett. **44**, 651 (1984)
3.4 E. Betzig, A. Lewis, A. Harootunian, M. Isaacson, E. Kratschmer: Biophys. J. **49**, 269 (1986)
3.5 T. Yatsui, M. Kourogi, M. Ohtsu: Appl. Phys. Lett. **73**, 2090 (1998)
3.6 T. Saiki, S. Mononobe, M. Ohtsu, N. Saito, J. Kusano: Appl. Phys. Lett. **68**, 2612 (1996)
3.7 S. Mononobe, T. Saiki, T. Suzuki, S. Koshihara, M. Ohtsu: Opt. Commun. **146**, 1299 (1998)
3.8 K. Kurihara, M. Ohtsu, T. Yoshida, T. Abe, H. Hisamoto, K. Suzuki: Anal. Chem. **71**, 3558 (1999)
3.9 T. Yatsui, M. Kourogi, K. Tsutsui, J. Takahashi, M. Ohtsu: Opt. Lett. **25**, 1279 (2000)
3.10 P.N. Minh, T. Ono, M. Esashi: Appl. Phys. Lett. **75**, 4076 (1999)
3.11 T. Matsumoto, T. Ichimura, T. Yatsui, M. Kourogi, T. Saiki, M. Ohtsu: Opt. Rev. **5**, 369 (1998)
3.12 U.M. Rajagopalan, S. Mononobe, J. Yoshida, M. Yoshimoto, M. Ohtsu: Jpn. J. Appl. Phys. **38**, 6713 (1999)

3.13 M. Naya, R. Micheletto, S. Mononobe, R. Uma Maheswari, M. Ohtsu: Appl. Opt. **36**, 1681 (1997)
3.14 T. Saiki, K. Nishi, M. Ohtsu: Jpn. J. Appl. Phys. **37**, 1638 (1998)
3.15 N. Hosaka, T. Saiki: J. Microscopy **202**, 362 (2001)
3.16 Y. Narita, T. Tadokoro, T. Ikeda, T. Saiki, S. Mononobe, M. Ohtsu: Appl. Spectroscopy **52**, 1141 (1998)
3.17 Y. Yamamoto, M. Kourogi, M. Ohtsu, V.V. Polonski, G.H. Lee: Appl. Phys. Lett. **76**, 2173 (2000)
3.18 V.V. Polonski, Y. Yamamoto, M. Kourogi, H. Fukuda, M. Ohtsu: J. Microscopy **194**, 545 (1999)
3.19 T. Kawazoe, Y. Yamamoto, M. Ohtsu: Appl. Phys. Lett. **79**, 1184 (2001)
3.20 I.I. Smolyaninov, D. Mazzoni, C.C. Davis: Appl. Phys. Lett. **67**, 3859 (1995)
3.21 K. Liberman, Y. Shani, I. Melnic, S.Yoffe, Y. Sharon: J. Microscopy **194**, 537 (1999)
3.22 M. Ohtsu: J. of Jpn. Soc. for Precision Eng. **58**, 410 (1992)
3.23 H. Ito, T. Nakata, K. Sakaki, M. Ohtsu, K.I. Lee, W. Jhe: Phys. Rev. Lett. **76**, 4500 (1996)
3.24 H. Hori: Oyo Buturi **68**, 180 (1999)
3.25 K. Cho, H. Ishihara, Y. Ohfuchi: Buturi **52**, 343 (1997)
3.26 O.J. Martin, C. Girard, A. Dreux: Phys. Rev. Lett. **74**, 526 (1995)
3.27 V.V. Klimov, V.S. Letokhov: Opt. Commun. **122**, 155 (1996)
3.28 Y. Sugarawa: Application of Mechanical Action (2): Surfaces. In: *Handbook of Near-Field Nanophotonics*, ed. by M. Ohtsu, S. Kawata (Optronics Publishers, Tokyo 1997) pp. 201–207

Chapter 4

4.1 I. Banno, H. Hori: Trans. IEE Japan, **119**-C, 1094 (1999)
4.2 M. Ohtsu (Ed.): *Near-Field Nano/Atom Optics and Technology*, (Springer-Verlag, Tokyo Berlin 1998), Chap. 2
4.3 M. Naya, S. Mononobe, R.U. Maheswari, T. Saiki, M. Ohtsu: Opt. Commun. **124**, 9 (1996)
4.4 A.V. Zvyagin, J.D. White, M. Ohtsu: Opt. Lett. **22**, 955 (1997)
4.5 K. Kobayashi, O. Watanuki: J. Vac. Sci. Technol. B **14**, 804 (1996)

Chapter 5

5.1 E.A. Hinds: Perturbative Cavity Quantum Electrodynamics. In: *Cavity Quantum Electrodynamics*, ed. by P.R. Berman (Academic Press, Boston 1994) pp. 1–56

Chapter 6

6.1 K. Kobayashi: Review of the Conventional Theories. In: *Handbook of Near-Field Nanophotonics*, ed. by M. Ohtsu, S. Kawata (Optronics Publishers, Tokyo 1997) pp. 233–239
6.2 M. Ohtsu (Ed.): *Near-Field Nano/Atom Optics and Technology* (Springer-Verlag, Tokyo Berlin 1998) Chap. 12
6.3 K. Kobayashi, O. Watanuki: J. Vac. Sci. Technol. B **15**, 1966 (1997)

Chapter 7

7.1 I. Banno, H. Hori: Jpn. J. Inst. Electric. Eng. **119**-C, 1094 (1999)
7.2 I. Banno, H. Hori: private communication (2000)
7.3 A. Ogiwara: *Light-Scattering from a Small Dielectric Sphere*, Thesis, Faculty of Engineering, Yamanashi University, Yamanashi (1997)
7.4 O.J.F. Martin, C. Girard, A. Dereux: Phys. Rev. Lett. **74**, 526 (1995)

Chapter 8

8.1 K. Kobayashi, S. Sangu, M. Ohtsu: Quantum Theoretical Approach to Optical Near-Fields and some Related Applications. In: *Progress in Nano-Electro-Optics I*, ed. by M. Ohtsu (Springer-Verlag, Berlin 2002) pp. 119–157

Chapter 9

9.1 M. Ohtsu (Ed.): *Near-Field Nano/Atom Optics and Technology* (Springer-Verlag, Tokyo Berlin 1998) Chap. 2
9.2 H. Kuhn: J. Chem. Phys. **53**, 101 (1970)
9.3 T. Förster: Delocalized Excitation and Excitation Transfer. In: *Modern Quantum Chemistry*, ed. by O. Sinanoğlu (Academic Press, New York 1965) pp. 93–138 (Sect. III-1)
9.4 N. Hosaka, T. Saiki: J. Microscopy **202**, 362 (2001)
9.5 N. Hosaka, T. Saiki: 8 nm Super Resolution on Near-Field Fluorescence Imaging of Single Molecules. In: *CLEO/QELS 2001* (2001) QPD7-1
9.6 X.S. Xie, R.C. Dunn: Science **265**, 361 (1994)
9.7 M. Ohtsu (Ed.): *Near-Field Nano/Atom Optics and Technology* (Springer-Verlag, Tokyo Berlin 1998) Chap. 11
9.8 V.I. Balykin, V.S. Letokhov, Yu.B. Ovchinnikov, A.I. Sidorov: Phys. Rev. Lett. **60**, 2137 (1988)
9.9 A. Landragin, J.-Y. Courtois, G. Labeyrie, N. Vansteenkiste, C.I. Westbrook, A. Aspect: Phys. Rev. Lett. **77**, 1464 (1996)
9.10 D. Marcuse: *Light Transmission Optics*, 2nd. edn. (Robert E. Krieger, Malabar 1989)
9.11 P.R. Berman (Ed.): *Cavity Quantum Electrodynamics* (Academic Press, San Diego 1994)
9.12 P.W. Milonni: *The Quantum Vacuum* (Academic Press, Boston 1994)
9.13 K. Kobayashi, S. Sangu, H. Ito, M. Ohtsu: Phys. Rev. A **63**, 013806 (2001)
9.14 R. Newton: *Scattering Theory of Waves and Particles*, 2nd. edn. (Dover Publications, New York 2002)
9.15 K. Kobayashi, S. Sangu, H. Ito, M. Ohtsu: Effective Probe–Sample Interaction: Toward Atom Deflection and Manipulation. In: *Near-Field Optics: Principles and Applications*, ed. by X. Zhu, M. Ohtsu (World Scientific, Singapore 2000) pp. 82–88
9.16 M. Ohtsu, K. Kobayashi, T. Kawazoe, S. Sangu, T. Yatsui: IEEE J. Sel. Top. Quant. Electron. **8**, 839 (2002)
9.17 T. Yatsui, M. Kourogi, M. Ohtsu: Appl. Phys. Lett. **79**, 4583 (2001)

9.18 T. Kawazoe, K. Kobayashi, J. Lim, Y. Narita, M. Ohtsu: Verification of Principle for Nanometer Size Optical Near-Field Switch by Using CuCl Quantum Cubes. In: *CLEO/Pacific Rim 2001* (2001) MH2-4
9.19 K. Kobayashi, T. Kawazoe, S. Sangu, J. Lim, M. Ohtsu: Theoretical and Experimental Study on a Near-Field Optical Nano-Switch, *IPR/PS 2001* (2001) PThB4
9.20 T. Kawazoe, K. Kobayashi, J. Lim, Y. Narita, M. Ohtsu: Phys. Rev. Lett. **88**, 067404 (2002)
9.21 K. Kobayashi, S. Sangu, T. Kawazoe, A. Shojiguchi, K. Kitahara and M. Ohtsu, J. Microsc. **210**, 247 (2003).
9.22 S. Sangu, K. Kobayashi, A. Shojiguchi, T. Kawazoe, and M. Ohtsu, J. Appl. Phys. **93**, 2937 (2003).
9.23 T. Kawazoe, K. Kobayashi, S. Sangu, and M. Ohtsu, Appl. Phys. Lett. **82**, 2957 (2003).

Appendix C

C.1 H. Haken: *Quantum Field Theory of Solids* (North-Holland, Amsterdam 1976)
C.2 J.J. Hopfield: Phys. Rev. **112**, 1555 (1958)
C.3 C. Kittel: *Quantum Theory of Solids*, 2nd revised printing (John Wiley, New York 1987)
C.4 D. Pines: *Elementary Excitation in Solids* (Perseus Books, Reading 1999)
C.5 P.W. Anderson: *Concepts in Solids* (World Scientific, Singapore 1997)

Appendix D

D.1 J.J. Sakurai: *Advanced Quantum Mechanics* (Addison-Wesley, Reading 1967)
D.2 A. Messiah: *Quantum Mechanics* (North-Holland, Amsterdam 1961)
D.3 D.P. Craig, T. Thirunamachandran: *Molecular Quantum Electrodynamics* (Dover Publications, New York 1998)
D.4 H. Haken: *Light* (North-Holland, Amsterdam 1986)
D.5 J.J. Hopfield: Phys. Rev. **112**, 1555 (1958)
D.6 J. Knoester, S. Mukamel: Phys. Rev. A **40**, 7065 (1989)
D.7 B. Huttner, S.M. Barnett: Phys. Rev. A **46**, 4306 (1992)
D.8 G. Juzeliūnas, D.L. Andrews: Phys. Rev. B **49**, 8751 (1994)
D.9 K. Kobayashi, S. Sangu, H. Ito, M. Ohtsu: Phys. Rev. A **63**, 013806 (2001)
D.10 H. Hyuga, H. Ohtsubo: Nucl. Phys. A **294**, 348 (1978)
D.11 P. Fulde: *Electron Correlations in Molecules and Solids*, 3rd edn. (Springer-Verlag, Berlin 1995)
D.12 K. Kobayashi, M. Ohtsu: J. Microsc. **194**, Part 2/3, 249 (1999)
D.13 M. Ohtsu, K. Kobayashi, H. Ito, G.H. Lee: Proc. IEEE **88**, 1499 (2000)
D.14 S. Sangu, K. Kobayashi, M. Ohtsu: J. Microsc. **202**, Part 2, 279 (2001)

Index